WHEN MONSTERS RULED

American Civil War
American History
American Politics
American South
Ancient History
Anthropology
Biblical Exegesis
Biblical Hermeneutics
Biography
Christian Mysticism
Coffee Table
Comparative Mythology
Comparative Religion
Cooking
Diet and Nutrition
Education
Ethnic Studies
Etymology
European History
Evolutionary Biology
Exposés
Family Histories
Film
Genealogy
Ghost Stories
Health and Fitness
Humor
Law of Attraction
Life After Death
Matriarchy
Men
Metaphysics

Military History
Mysteries and Enigmas
Natural Health
Natural History
Onomastics
Paleography
Paleontology
Paranormal
Patriarchy
Philosophy
Photography
Poetry
Politics
Presidential History
Quiz
Reference
Religion
Revolutionary Period
Science
Self-help
Spirituality
Spiritualism
Sport Science
Technology
Thanatology
Thealogy
Theology
UFOlogy
Victorian Period
Wildlife
Women
World History

*Mr. Seabrook does not author books for fame and glory,
but for the love of writing and sharing his knowledge.*

SeaRavenPress.com

Warning: SEA RAVEN PRESS BOOKS WILL EXPAND YOUR ★ MIND!

When

MONSTERS

Ruled

THE 25 SCARIEST ANIMALS OF THE PREHISTORIC WORLD

------ ◆◦{B◦B}◦◆ ------

LOCHLAINN SEABROOK

Bestselling Author, Award-winning Historian, Acclaimed Artist

Diligently Researched and Generously Illustrated
by the Author for the Elucidation of the Reader

2025

Sea Raven Press, Park County, Wyoming, USA

WHEN MONSTERS RULED

Published by
Sea Raven Press, LLC, Founded 1995
Cassidy Ravensdale, President
Park County, Wyoming, USA
SeaRavenPress.com

SEA RAVEN PRESS
ARTISAN-CRAFTED BOOKS & MERCH FROM THE ROCKY MOUNTAINS

PRINTING HISTORY
1st SRP paperback edition, 1st printing, June 2025 • ISBN: 978-1-955351-58-4
1st SRP hardcover edition, 1st printing, June 2025 • ISBN: 978-1-955351-59-1

ISBN: 978-1-955351-58-4 (paperback)
Library of Congress Control Number: 2025939491

When Monsters Ruled: The 25 Scariest Animals of the Prehistoric World, by Lochlainn Seabrook. Includes an introduction, illustrations, index, and bibliography.

WARNING: THE MATERIAL IN THIS BOOK MAY NOT BE SUITABLE FOR YOUNG CHILDREN. RECOMMENDED AGES: 10 AND UP.

ARTWORK
Front and back cover design and art, book design, layout, font selection, and interior art by Lochlainn Seabrook.
All images, pictures, photos, illustrations, image captions, graphic design, and graphic art copyright © Lochlainn Seabrook.
All images selected, placed, manipulated, cleaned, colored, tinted, and/or created by Lochlainn Seabrook.
All rights reserved.

The views documented in this book concerning prehistoric life and the prehistoric world are those of the publisher.

WRITTEN, DESIGNED, PUBLISHED, PRINTED, & MANUFACTURED IN THE UNITED STATES OF AMERICA

PALEONTOLOGY MATTERS

DEDICATION

To all those who devote their lives to the study of the past.

EPIGRAPH

"Thus the great drama of universal life is perpetually sustained; and though the individual actors undergo continual change, the same parts are ever filled by another and another generation; renewing the face of the earth, and the bosom of the deep, with endless successions of life and happiness."

Richard Buckland, 1836

PALEONTOLOGIST, GEOLOGIST, THEOLOGIAN

CONTENTS

Allosaurus fragilis. Copyright © Lochlainn Seabrook.

Ceratosaurus. Copyright © Lochlainn Seabrook.

NOTES TO THE READER

☞ Due to the often utter scarcity of fossils of prehistoric animals, including many of those listed in this book, my illustrations of an individual animal may differ slightly from one another. In fact, this is to be expected. In many cases, take Gigantopithecus for instance, we currently lack even postcranial bones, and must construct a physical image of the creature with only a few teeth and mandibles. Thus a convincing theoretical picture must be built up using comparative anatomy—the result being that images of the same animal may differ from picture to picture.

☞ In addition to the foregoing comments, in illustrating this book I have granted myself certain artistic liberties, in some cases for the sake of dramatic effect. Thus, not all images are to scale and therefore should not be regarded as precisely accurate. However, my illustrations do reflect most of the commonly accepted basic current knowledge and theories we have regarding all 25 animals. Pursuing the mysteries of science is never ending.

L.S.

A pack of *Velociraptor mongoliensis* on the hunt. Copyright © Lochlainn Seabrook.

A 40 foot long, 5.5 ton *Saurophaganax maximus*. Copyright © Lochlainn Seabrook.

INTRODUCTION

Welcome to the amazing paleontological world of prehistoric animals, a realm so fantastical that few would believe it ever existed—at least not without scientific evidence to prove it. Fortunately for us, hundreds of thousands of fossils have been collected over the centuries, all bearing mute but irrefutable testimony to the fact that this often seemingly preposterous ancient world, one filled with unaccountable monsters of all kinds, did indeed endure for tens of millions of years. For those who are interested in this time-forgotten era I have penned *When Monsters Ruled*.

In the following pages you will encounter some of the most terrifying beasts to have ever walked our planet; living nightmares such as Mosasaurus, Gigantopithecus, Titanoboa, megalodon, and of course, Tyrannosaurus. What do all of these animals have in common? As massive apex predators all once reigned supreme over their respective habitats and time periods. Many, such as the Allosaurids, survived for a staggering 55 million years. Yet, now all are gone, casualties of violent ecological forces far beyond their limited understanding and control.

This makes us pause to reflect on a frightening scenario: What if some of them had lived into the present? What would our lives be like today? I am speaking in some cases of carnivorous terrors that weighed 50 tons, sported 12 inch teeth, were as big as school buses, and swallowed animals the size of cars—whole.

Read on and discover the ecology and life history of these extraordinary and often inexplicable primordial species for yourself. Contemplate what it would have been like to exist at the same time.

(Humans actually lived alongside some of them!) When all is said and done, you will come away both impressed and beguiled by the many amazing traits, capabilities, and behaviors of these paleofauna—all products of millions of years of evolutionary forces, set in motion by God/Nature. A truly unimaginable and breathtaking world awaits you.

Lochlainn Seabrook
Park County, Wyoming, USA
June 2025
Lux est Veritas

SEA RAVEN PRESS
PARK COUNTY ❦ WYOMING USA
EST. 1995

"Books invite all; they constrain none."
Hartley Burr Alexander (1873-1939)

WHEN MONSTERS RULED

1
ANDREWSARCHUS

Andrewsarchus mongoliensis. Copyright © Lochlainn Seabrook.

SCIENTIFIC DETAILS

COMMON NAME: Andrewsarchus.

SCIENTIFIC NAME: *Andrewsarchus mongoliensis.*

ETYMOLOGY: *Andrews* derives from the name of its discoverer, Roy Chapman Andrews; *archus* is from the Greek word *archos*, "ruler"; *Mongoliensis* refers to Mongolia, where its remains were first found. Full meaning: "Andrew's ruler from Mongolia."

NICKNAME: "The giant cetancodontamorph of Mongolia."

CLASS: Mammal.

SHOULDER HEIGHT: Up to 6 feet 6 inches.

LENGTH: Possibly up to 16 feet.

WEIGHT: Possibly up to 2,200 lbs (1.1 tons).

HABITAT: Temperate and subtropical regions.

Andrewsarchus. Copyright © Lochlainn Seabrook.

LIFE SPAN: Probably about 30 years.

TIME PERIOD: Middle Eocene, 45 to 36 million years ago.

DIETARY CATEGORY: Probably an omnivore.

DIET: Meat, carrion, vegetation and plant matter.

HUNTING METHODS: Predatory stalk and ambush.

TOP SPEED: Possibly 5 mph for short distances.

GEOGRAPHIC RANGE: Coastal areas of central Asia.

REASONS FOR EXTINCTION: Climate change, geological changes, loss of prey species, competition with other animals.

DESCRIPTION

Skeleton of *Andrewsarchus*. Copyright © Lochlainn Seabrook.

Discovered by Andrews' team in Mongolia in 1923, *Andrewsarchus* is believed by some to be the largest predatory mammal to have ever lived on our planet. Its massive skull, of which there is currently only one specimen, was nearly 33 inches in length.

Due to the fact that only a few remains (no leg bones) have been unearthed thus far, its exact appearance and scientific classification cannot yet be fully or accurately understood. We do know that it was an artiodactyl (an even-toed ungulate), and that it was a cetancodontamorph; that is, a close relation of hippopotamuses, whales, and dolphins.

Its large pointed canine teeth would seem to give bold evidence of its life as a terrifying mammalian killer with an insatiable carnivorous appetite. However, its non-canine teeth show a lack of specialization, indicating that it is more likely to have been an omnivore, one that intermittently hunted live prey, scavenged on carcasses, and, when necessary, fed on assorted vegetation. Carnivore or omnivore, its powerful jaws, massive shoulder muscles, and speed would have made it an intimidating opponent.

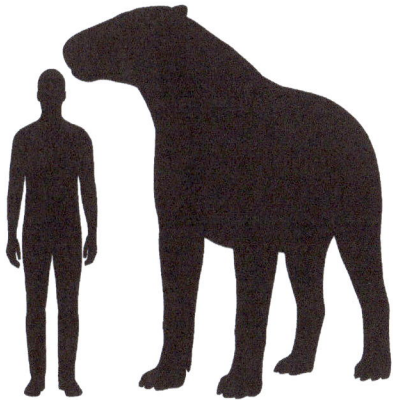

Size comparison between *Andrewsrarchus* and man. Copyright © Lochlainn Seabrook.

Many aspects of *Andrewsarchus* remain a mystery, riddles that will only be solved when more fossils are one day uncovered.

Andrewsarchus mongoliensis hunting a 4 ton male *Megacerops*. Copyright © Lochlainn Seabrook.

2
ARCTODUS

Arctodus simus. Copyright © Lochlainn Seabrook.

SCIENTIFIC DETAILS

COMMON NAME: Giant short-faced bear.

SCIENTIFIC NAME: *Arctodus simus*.

ETYMOLOGY: *Arcto* is Greek for "bear"; *odus* is Greek for "tooth"; and *simus* is Greek for "snub-nosed." Full meaning: "Bear tooth blunt-nosed."

NICKNAME: "Bulldog bear."

CLASS: Mammal.

SHOULDER HEIGHT: 5.9 feet.

STANDING HEIGHT: Up to 20 feet.

LENGTH: Up to 12 feet.

WEIGHT: Over 3,500 lbs (1.75 tons).

Arctodus. Copyright © Lochlainn Seabrook.

HABITAT: Savannah, steppe, open plains, tundra, grassland.

LIFE SPAN: Possibly 25 years.

TIME PERIOD: Pleistocene, 2.6 million to 11,700 years ago.

DIETARY CATEGORY: Carnivore, but possibly somewhat omnivorous, thus also consuming roots, nuts, and fruits.

DIET: Mammoths, bison, ground sloths, horses, camels, caribou, deer, carrion.

HUNTING METHODS: Predatory, scavenging, ambush hunting, stalking, chasing, and probably kleptoparasitism.

TOP SPEED: Up to 40 mph for short bursts.

GEOGRAPHIC RANGE: North America, Alaska to central Mexico.

DESCRIPTION

Considered by some to have been the largest, or at least one of the largest, terrestrial carnivores in world history, the giant short-faced bear was an immense long-legged predator that terrorized prehistoric North America for thousands of millennia. One scientist has rightfully described *Arctodus* as without question the most powerful Pleistocene predator to have stalked the American continent. In fact, its sheer size, speed, power, and

Skeleton of *Arctodus* Copyright © Lochlainn Seabrook.

ferocity may be one of the primary reasons that humans settled the interior of North America later than South America: It was almost impossible for our ancestors to survive an encounter with this intimidating colossal beast, one whose front legs alone spanned a remarkable 9 feet.

Closely related to modern spectacled bears (*Tremarctos ornatus*), this frightening Ice Age ursid was probably solitary, denned in caves, and wandered over enormous distances in search of sustenance, no doubt following the migratory patterns of his favorite food: herbivorous mammals.

North America's largest carnivore, far bigger than even modern day polar bears, it possessed great intelligence, cursorial limbs, dagger like canines, huge bone-crushing jaws, enormous 6 inch claws, and superlative speed, all which helped it survive for some 2 million years before succumbing to extinction. The causes for its demise were probably the same as those that killed off most other Ice Age megafauna: Competition with newly evolving species, human hunting, prey decline, a rigid diet, and climate change.

Size comparison between *Arctodus* and man. Copyright © Lochlainn Seabrook.

Arctodus simus attacking a 2 ton male steppe bison. Copyright © Lochlainn Seabrook.

3
ARGENTAVIS

Argentavis magnificens. Copyright © Lochlainn Seabrook.

SCIENTIFIC DETAILS

COMMON NAME: Argentavis.

SCIENTIFIC NAME: *Argentavis magnificens*.

ETYMOLOGY: *Argent* derives from Argentina, where it was first found; *avis* is the Latin word for "bird"; and *magnificens* is Latin for "magnificent." Full meaning: "Magnificent bird from Argentina."

NICKNAMES: "Giant teratorn," "Thunderbird."

CLASS: Aves (birds).

HEIGHT: Up to 6 feet.

WINGSPAN: By some estimates up to 30 feet.

WEIGHT: As much as 200 lbs.

HABITAT: Mountains, plains, scrubland.

LIFE SPAN: Possibly up to 100 years.

TIME PERIOD: Late Miocene Epoch, 6 to 8 million years ago.

Head of *Argentavis*. Copyright © Lochlainn Seabrook.

DIETARY CATEGORY: Carnivore.

DIET: Carrion, possibly small game: rodents, reptiles, other birds.

HUNTING METHOD: Mainly scavenging, but also possibly opportunistic predation.

TOP SPEED: Up to 50 mph.

GEOGRAPHIC RANGE: Argentina.

REASONS FOR EXTINCTION: Climate change, habitat reduction, loss of megafauna upon which to feed, ever increasing competition from newly emerging predators and scavengers.

DESCRIPTION

Skeleton of *Argentavis*. Copyright © Lochlainn Seabrook.

With a probable wingspan of around 30 feet, a height of 6 feet, and a body length (beak to tail) of 11 feet, *Argentavis* was the largest known flying bird ever to inhabit our skies. For comparison it was the same size as a modern Piper PA-28 Cherokee 140, a popular light aircraft with a 30 foot wingspan that is often used as an air taxi. Today's largest living bird is the wandering albatross (*Diomedea exulans*), with a wingspan of around 11 feet. *Argentavis* was nearly three times bigger.

Its fossilized bones reveal a heavily built skeleton with a robust breastbone for large muscle attachment—evidence that it was powerful flyer. However, like modern vultures, it was probably not capable of long distance "flapping" flight. More likely it was an elegant soaring and gliding bird, a living "sailplane" that used updrafts and thermal columns to sustain itself in the air for hours at a time.

While *Argentavis* was a carnivorous predator equipped with 4 inch talons and a meat cleaver beak, it was not a true raptor like modern eagles. Instead, it was probably a scavenger, one that used its fierce looking claws for both perching and tearing apart the meat of dead animals. We cannot rule out, however, that it occasionally hunted small game when necessary.

One of the more fascinating aspects of the giant teratorn is that it has long been associated with the Thunderbird, a massive dark-feathered avian that, according to Native American mythology, is said to rule the skies into the present day. Indeed, sightings have been consistently reported by credible eyewitnesses across the U.S., all which perfectly match the

Argentavis was the size of a small airplane. Copyright © Lochlainn Seabrook.

description of *Argentavis magnificens*. Is this a case, like the coelacanth, of a Lazaurus taxon, an animal once believed to be extinct but which is actually still very much alive?

Argentavis magnificens feeding on carrion. Copyright © Lochlainn Seabrook.

4
ARTHROPLEURA

Arthropleura armata. Copyright © Lochlainn Seabrook.

SCIENTIFIC DETAILS

COMMON NAME: Giant millipede.

SCIENTIFIC NAME: *Arthropleura armata*.

ETYMOLOGY: *Arthro* is Greek for "joint"; *pleura* is Greek for "side" plates; and *armata* is Latin for "armored." Full meaning: "armored joint-sided."

NICKNAME: "Carboniferous giant."

CLASS: Diplopoda (arthropodic millipedes).

WIDTH: 22 inches.

LENGTH: 8.5 feet.

WEIGHT: 110 lbs.

Head of *Arthropleura*. Copyright © Lochlainn Seabrook.

HABITAT: Tropical forests, steamy swamps, ocean shorelines.

LIFE SPAN: Possibly 10 to 15 years.

TIME PERIOD: Carboniferous, 345 to 280 million years ago.

DIETARY CATEGORY: Most likely an herbivore, or more specifically an opportunistic detritivore; not likely carnivorous.

DIET: Seeds, wood, leaves, forest litter, and decaying plant matter; if carnivorous, however, invertebrates such as worms, insects, spiders, larvae, and small amphibious creatures.

HUNTING METHOD: Probably a harmless plant-eater.

TOP SPEED: Slow-moving; possibly up to 1.2 mph.

GEOGRAPHIC RANGE: Europe and North America.

REASONS FOR EXTINCTION: Climate change, decreased oxygen levels, dietary alterations, ascendence of vertebrates.

DESCRIPTION

First discovered in Germany in 1854 amid shale deposits and coal seams, this gigantic, nearly 9 foot long centipede-like creature holds the record for being the largest terrestrial invertebrate to have ever lived. Vertebrate animals had not yet begun to dominant the landscape, thus it was able to grow to a tremendous size—the same length as a standard pool table, surfboard, or adult male alligator. It was, in fact, 3 feet longer than the height of an average adult North American woman.

Trackway of *Arthropleura armata*. Copyright © Lochlainn Seabrook.

Like *Andrewsarchus*, we do not yet have a complete fossil of *Arthropleura*, making it impossible to accurately know and understand its morphology. However, its fossilized trackways, showing evidence of numerous pairs of short legs, provide an idea of how it moved.

From fossil remains it appears that *Arthropleura's* body was comprised of 30 jointed segments, all which were covered by two lateral main plates and one central plate. Its two short trumpet-like antennae were offset by two bulbous eyes on tiny stalks and a small mouth suited to fragmentizing and processing vegetable matter. In order to grow in size, like its modern day relatives the crustaceans, arachnids, centipedes, and millipedes, *Arthropleura* probably molted its inflexible exoskeleton several times a year.

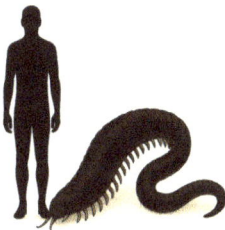

Size comparison between *Arthropleura* and man. Copyright © Lochlainn Seabrook.

A huge prehistoric arthropod with a heavily armored exoskeleton and a flattened segmented body, due to its immense size it probably had few enemies; although it was almost certainly occasionally preyed upon by various species of primitive amphibians and reptiles, especially when young and also during molting.

Despite its terrifying aspects, scientifically speaking *Arthropleura armata* is an amazing example of giant myriapod evolution. Fortunately for us this massive beast went extinct 280 million years ago.

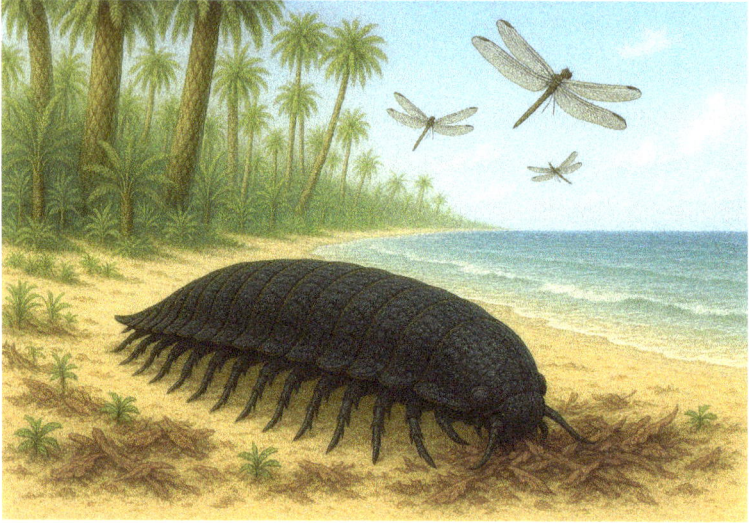

Arthropleura feeding on beach detritus with giant dragon flies (with 2.5 feet wingspans) in the background. Copyright © Lochlainn Seabrook.

Arthropleura armata in its natural habitat: a humid Carboniferous forest. Copyright © Lochlainn Seabrook.

5
BASILOSAURUS

Basilosaurus caucasicus. Copyright © Lochlainn Seabrook.

SCIENTIFIC DETAILS

COMMON NAME: Basilosaurus.

SCIENTIFIC NAME: *Basilosaurus caucasicus*.

ETYMOLOGY: *Basileus* is Greek for "king"; *saurus* is Greek for "lizard"; and *caucasicus* is a Latinization of the word Caucasus, a mountainous region near the Black Sea. Full meaning: "King lizard from the Caucasus."

NICKNAMES: "Serpent whale," "sea serpent of the Eocene."

CLASS: Mammal.

HEIGHT: Possibly 6 feet.

LENGTH: Up to 60 feet.

Basilosaurus breaching. Copyright © Lochlainn Seabrook.

WEIGHT: Up to 33,000 lbs (16.5 tons).

HABITAT: Shallow tropical and subtropical seas.

LIFE SPAN: Possibly 40 to 50 years.

TIME PERIOD: Late Eocene epoch, 40 to 34 million years ago.

DIETARY CATEGORY: Carnivore.

DIET: Fish, squid, other marine mammals.

HUNTING METHOD: Ambush predation.

TOP SPEED: Slow, perhaps 5 to 8 mph, with short 12 mph bursts.

GEOGRAPHIC RANGE: The tropical Tethys Ocean; remains found in Russia, Eastern Europe, Egypt, and the United States.

REASONS FOR EXTINCTION: Climate change, outcompeted by modern more highly developed whale species, declining prey species, habitat loss.

DESCRIPTION

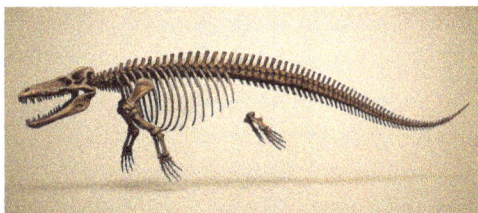

Skeleton of *Basilosaurus*. Copyright © Lochlainn Seabrook.

First discovered in Russia in 1934, this terrifying, prehistoric, reptile-like whale ruled the warm coastal shallows of the Tethys Ocean for some 4 million years. Its lean snake-like body, massive teeth-filled jaws, extra long skull, and probable lateral (as opposed to vertical) tail fluke gave it a bizarre appearance that sets it apart from modern whales. Also, unlike many of today's cetaceans, which are filter-feeders, *Basilosaurus* was more similar to our orcas (killer whales), making it a predatory creature that fed on fish, cephalopods, and other marine mammals.

A true "sea monster" in every sense, this stream-lined, primitive apex predator, with its elongated vertebrae and serpentine body, seems to have possessed an anguilliform morphology; that is, it moved in an eerie undulating eel-like manner. Its worm-like flexibility, along with its keen underwater hearing, allowed it to stealthily sneak up on its prey from the dark depths below before launching a sudden violent attack on its unsuspecting victim.

Though unuseable, *Basilosaurus* had vestigial rear limbs, showing that it was a transitional whale form, evidence of the evolutionary processes that gradually turned this former air-breathing land mammal into an air-breathing aquatic mammal. The fossil record is not yet clear on how its tail looked. While some postulate a long tapered reptilian-like tail, others assume it may have had a small rudimentary fluke at

Size comparison between *Basilosaurus* and man. Copyright © Lochlainn Seabrook.

the end, used, not for thrust and propulsion, as in modern whales, but as an aid in steering, braking, and stabilization.

Basilosaurus disappeared with the arrival of modern more highly specialized whale species, dramatic changes in the temperature and weather, and loss of prey and habitat.

Basilosaurus hunting a *Dorudon atrox*. Copyright © Lochlainn Seabrook.

Basilosaurus caucasicus attacking a squid. Copyright © Lochlainn Seabrook.

6
BRYGMOPHYSETER

Brygmophyseter shigensis. Copyright © Lochlainn Seabrook.

SCIENTIFIC DETAILS

COMMON NAME: Brygmophyseter.

SCIENTIFIC NAME: *Brygmophyseter shigensis.*

ETYMOLOGY: *Brygmos* is Greek for "gnashing"; *physeter* is Greek for "blower"; and *shigensis* is a Latinized word meaning "from Shiga"—Shiga-mura, Japan, being the village where the whale's fossil was discovered. Full meaning: "The biting whale with a blowhole from Shiga, Japan."

NICKNAME: "The gnashing sperm whale."

CLASS: Mammal.

HEIGHT: 4 feet tall at the widest part.

The eye of *Brygmophyseter shigensis.*
Copyright © Lochlainn Seabrook.

LENGTH: About 23 feet.

WEIGHT: 10,000 lbs (5 tons).

HABITAT: Deep open oceans with continental slopes.

LIFE SPAN: Possibly 50 years.

TIME PERIOD: Middle Miocene, 14 to 15 million years ago.

DIETARY CATEGORY: Carnivore.

DIET: Other sea mammals, like whales and dolphins, as well as squid and fish—including sharks.

HUNTING METHOD: Raptorial feeding strategy.

TOP SPEED: Possibly 30 mph.

GEOGRAPHIC RANGE: Northwest Pacific region.

REASONS FOR EXTINCTION: Competition, climate change.

DESCRIPTION

Mouth of *Brygmophyseter*. Copyright © Lochlainn Seabrook.

Discovered in Japan in the 1980s, *Brygmophyseter* was a prehistoric toothed cetacean and a member of the sperm whale family. Like its modern cousins, it hunted in pods using echo-location pulses that could both detect and stun its prey. Members of the pod protected one another using their enormous heads to ram and kill would-be attackers.

This apex predator possessed huge, imposing, sharp, peg-like teeth. However, unlike modern sperm whales, which have teeth only in their lower jaws, *Brygmophyseter* had teeth in both its upper and lower jaws, helping it grasp and retain writhing slippery prey.

It was a member of the macroraptorials, a fearsome group that includes other extinct large-toothed sperm whales, all known for violently hunting and feeding on massive marine animals. As such, it filled an ecological role similar to that of today's orcas.

Due to its size, innate weaponry, and aggressiveness, *Brygmophyseter* had few natural

Brygmophyseter skeleton. Copyright © Lochlainn Seabrook.

enemies, allowing it to grow slowly over many decades, not reaching full maturity until around 15 years of age. Perhaps like its macroraptorial relations, this over-specialized cetacean died out due to a series of intricate factors, including being outcompeted by newly evolving whale species, climate change, cooling ocean temperatures, loss of prey species, and

Size comparison between *Brygmophyseter* and man. Copyright © Lochlainn Seabrook..

dropping sea levels. In other words, it could not adapt quickly enough to the rapid changes developing around it. Today's oceans would be far different had it survived.

Brygmophyseter shigensis hunting a 43 foot long giant squid. Copyright © Lochlainn Seabrook.

7
CAMEROCERAS

Cameroceras trentonense. Copyright © Lochlainn Seabrook.

SCIENTIFIC DETAILS

COMMON NAME: Cameroceras.

SCIENTIFIC NAME: *Cameroceras trentonense.*

ETYMOLOGY: *Camero* is Latin for "chambered"; *ceras* is ancient Greek for "horn"; and *trentonense* is a Latinized word meaning "from Trenton," referring to the limestone formations known as the Trenton Group, located in New York, where the creature was first discovered. Full meaning: "The chambered horn from Trenton."

NICKNAME: "The giant Ordovician cephalopod."

CLASS: Cephalopoda.

SHELL HEIGHT: 24 inches at thickest part.

LENGTH: 20 feet—possibly up to 30 feet.

WEIGHT: Up to 2,200 lbs (1.1 tons).

Beak of *Cameroceras*. Copyright © Lochlainn Seabrook.

HABITAT: Shallow tropical seas with reef ecosystems.

LIFE SPAN: Possibly up to 25 years.

TIME PERIOD: Ordovician period, 470 to 444 million years ago.

DIETARY CATEGORY: Carnivore.

DIET: Mollusks, trilobites, early primitive fish, other cephalopods.

HUNTING METHODS: Predatory ambush: stalk and pounce.

TOP SPEED: Slow heavy mover; 2.5 mph in short bursts.

GEOGRAPHIC RANGE: North America, Europe, Asia.

REASONS FOR EXTINCTION: Died during the Late Ordovician mass extinction event, about 444 million years ago, which dramatically altered both the climate and the oceans.

DESCRIPTION

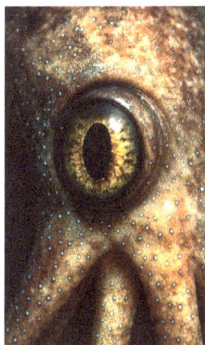

Cameroceras' eye. Copyright © Lochlainn Seabrook.

First discovered in 1842 in western New York state, *Cameroceras* possessed a long, dense, rigid, external, possibly 25 to 30 foot nautiloid shell which—similar to an actual nautilus shell—contained numerous inner chambers or compartments, allowing it to use fluids and gases to regulate its buoyancy.

Soft tissues do not usually preserve as fossils, so we do not know what the cephalic region looked like, but based on other similar straight shelled cephalopods we can surmise that it had an enormous head, long tentacles, and a large, powerful parrot-like beak for grasping, tearing, eating, and processing its prey: other aquatic animals.

This strange gigantic creature was one of the largest orthoconic (straight-shelled) cephalopods of the period and may be considered a true apex predator. Due to its enormous size it was no doubt slow-moving, however, making it most likely to have been an ambush predator rather than a pursuit predator.

Though it was one of the largest and most robust marine predators of its time, *Cameroceras* could not adapt to the effects of decreasing marine oxygen levels, drastic changes in sea level, global cooling, and other major ecological alterations—such as the evolution of more mobile coiled

Fossil shell of *Cameroceras*. Copyright © Lochlainn Seabrook.

cephalopodic forms. Thus, *Cameroceras*, along with all other straight-shelled species (such as *Endoceras* and *Orthoceras*), succumbed to the Late Ordovician mass extinction event that occurred some 444 million years ago.

Cameroceras' closest living relative is the beautiful but slowly disappearing *Nautilus pompilius*, a small coiled cephalopod known more commonly as the chambered nautilus.

Living chambered nautilus. Copyright © Lochlainn Seabrook.

Cameroceras trentonense attacking and eating a trilobite in a 450 million year old Ordovician ocean. Copyright © Lochlainn Seabrook.

8
CARBONEMYS

Carbonemys cofrinii. Copyright © Lochlainn Seabrook.

SCIENTIFIC DETAILS

COMMON NAME: Carbonemys.

SCIENTIFIC NAME: *Carbonemys cofrinii.*

ETYMOLOGY: *Carbo* is Latin for "coal," a reference to where it was discovered; *emys* is Greek for "freshwater turtle"; and *cofrinii* comes from the name of Victor Cofrini, a supporter of the fieldwork that led to this animal's discovery. Full meaning: "Cofrini's freshwater coal turtle."

NICKNAME: "The giant coal turtle."

CLASS: Reptile.

HEIGHT: 3 feet 3 inches.

LENGTH: 5 feet 6 inches.

WEIGHT: Possibly 1,545 lbs (0.77 tons).

Head of *Carbonemys.* Copyright © Lochlainn Seabrook.

HABITAT: Humid, dense tropical rainforests, coal-forming swamps, steamy lakes, slow-moving equatorial river systems.

LIFE SPAN: At least 100 years.

TIME PERIOD: Middle to Late Paleocene, 60 to 58 million years ago.

DIETARY CATEGORY: Carnivorous-oriented omnivore.

DIET: Mollusks, fish, reptiles, amphibians, crustaceans, carrion.

HUNTING METHOD: Ambush predation in freshwater systems.

TOP SPEED: Slow on land, 0.5 mph; in water about 3 mph.

GEOGRAPHIC RANGE: Northern South America.

REASONS FOR EXTINCTION: Climate change, new species.

DESCRIPTION

F irst discovered in the early 2000s in the coal-rich Cerrejón Formation of northeastern Columbia, South America, this reptilian behemoth was a meat-eating opportunistic omnivore who probably dined on almost anything it could catch or scavenge, from fish, frogs, amphibians, and crustaceans, to clams, reptiles, snails, and rotting meat.

Carbonemys in its natural habitat.
Copyright © Lochlainn Seabrook.

Like modern day snapping turtles, *Carbonemys*' powerful, durophagic, beak-like jaws allowed it to easily crush the hard shells and bones of its ambushed prey. When meat was not available it may have consumed aquatic plant matter to sustain itself. While little was safe around the mammoth predator, with few natural enemies itself it probably lived for a century or more.

As is common with many long-lived animals, this enormous, nearly 6 foot long freshwater turtle probably had a long maturation period and a slow reproductive cycle. It was at home in hot, rainy, humid tropical ecosystems, with fetid marshes, steamy jungles, placid lakes, and warm lazy rivers. Here it could hunt, feed, bask, and rest in muggy palm forests surrounded by equatorial ferns, fermenting tropical detritus, and humidity-soaked deadfall.

Its inevitable extinction came following the Paleocene-Eocene Thermal Maximum, which occurred about 56 million years ago. During this ecologically tumultuous 150,000 year period volcanic activity increased and carbon and methane were released in abnormally high quantities. The result was a worldwide greenhouse effect: the polar caps disappeared, temperatures increased, sea levels rose, and tropical ecosystems were dramatically disrupted. Additionally, more highly specialized turtles and crocodilians began to emerge, outcompeting *Carbonemys* in the struggle over ever shrinking resources. Though it survived for an astonishing 2 million years, today its fossils alone remain to tell the tale of this impressive but frightening reptile.

Size comparison between *Carbonemys* and man.
Copyright © Lochlainn Seabrook.

Carbonemys cofrinii feeding on one of its favorite foods: crustaceans. Copyright © Lochlainn Seabrook.

9

DEINOSUCHUS

Deinosuchus hatcheri. Copyright © Lochlainn Seabrook.

SCIENTIFIC DETAILS

COMMON NAME: Deinosuchus.

SCIENTIFIC NAME: *Deinosuchus hatcheri*.

ETYMOLOGY: *Deinos* is Greek for "terrible"; *suchus* is a Latinization of the Greek word *souchos*—the Greeks' rendering of the name of the Egyptian crocodile god Sobek; and *hatcheri* honors the name of American paleontologist John Bell Hatcher. Full meaning: "Hatcher's terrible crocodile."

NICKNAME: "The dinosaur-eating crocodile."

CLASS: Reptile.

SHOULDER HEIGHT: 5 feet.

LENGTH: Possibly up to 40 feet.

WEIGHT: Up to 10,000 lbs (5 tons).

Head of *Deinosuchus*. Copyright © Lochlainn Seabrook.

HABITAT: Swamps, shallow seas, estuaries, rivers, shorelines.

LIFE SPAN: Possibly 75 years or more.

TIME PERIOD: Late Cretaceous Period, 82 to 73 million years ago.

DIETARY CATEGORY: Carnivore.

DIET: Reptiles, fish, turtles, dinosaurs.

HUNTING METHODS: Ambush predator, scavenger.

TOP SPEED: On land 10 mph; in water 25 mph.

GEOGRAPHIC RANGE: North America.

REASONS FOR EXTINCTION: Climate change, competition with new species, habitat loss, and reduction of prey species.

DESCRIPTION

This gigantic 40 foot long crocodilian was first discovered in Montana in 1909, with additional fossil specimens later found in North Carolina, Georgia, Mississippi, Utah, Texas, and Mexico. One of the world's largest and most terrifying apex marine predators, its

Skeleton of *Deinosuchus hatcheri*. Copyright © Lochlainn Seabrook.

powerful jaws and massive conical teeth could deliver a bite force of over 20,000 psi—one of the strongest known in the animal kingdom.

Its awesome dimensions and strength meant that it had few if any natural enemies, allowing it to grow to an enormous size and live for nearly a century. Lurking along the coastlines of prehistoric North American bodies of water, like many other aquatic reptilian predators it used a grab-and-drown hunting method, crushing and breaking the bones of its prey, then dragging it out to deeper water to dismember and devour it. Even many of the largest dinosaurs had little hope of escape once they were targeted by *Deinosuchus*.

Size comparison between *Deinosuchus* and man. Copyright © Lochlainn Seabrook.

Though it was a member of the crocodilian order, it belonged to the alligator family, making it more closely related to the latter than the former. This frightening 5 ton beast met its end due to a major change in climate and the many alterations that accompanied it: newer more specialized marine reptiles, changes in temperature, and habitat and prey loss. We can only be thankful that *Deinsosuchus* did not live alongside us.

Deinosuchus hatcheri consuming an *Albertosaurus*. Copyright © Lochlainn Seabrook.

10
DUNKLEOSTEUS

Dunkleosteus terrelli. Copyright © Lochlainn Seabrook.

SCIENTIFIC DETAILS

COMMON NAME: Dunkleosteus.

SCIENTIFIC NAME: *Dunkleosteus terrelli*.

ETYMOLOGY: *Dunkle* honors paleontologist David Dunkle; *osteus* is Greco-Latin for "bone"; and *terrelli* honors the discoverer of the first fossil, Jay Terrelli. Full meaning: "Dunkle and Terrelli's bone."

NICKNAME: "Devil fish of the Devonian."

CLASS: Placodermia (extinct armored fishes).

HEIGHT: 6.5 feet at widest part.

LENGTH: Possibly up to 33 feet.

WEIGHT: 4,000 lbs to 8,000 lbs (2 to 4 tons).

HABITAT: Mid-level to deep water in shallow epicontinental seas.

Head of *Dunkleosteus*. Copyright © Lochlainn Seabrook.

LIFE SPAN: Possibly up to 25 years.

TIME PERIOD: Late Devonian, 382 to 358 million years ago.

DIETARY CATEGORY: Carnivore, apex predator, scavenger.

DIET: Fish, including sharks and other placoderms.

HUNTING METHOD: Ambush predation using mouth suction.

TOP SPEED: Probably 11 mph.

GEOGRAPHIC RANGE: North America, Europe, Northern Africa.

REASONS FOR EXTINCTION: Climate change, global cooling, collapse of food chain, changes in sea levels, oceanic anoxia, volcanic activity, possible asteroid impact.

DESCRIPTION

Probable *Dunkleosteus* skeleton. Copyright © Lochlainn Seabrook.

Dunkleosteus was first discovered in 1867 by Jay Terrelli in Ohio's Cleveland Shale Formation. Though a full skeleton has yet to be found, enough cranial and thoracic fossils have been gathered to develop a fairly accurate reconstruction of this formidable leviathan of the deep. Yet its true length and weight are still purely conjectural.

Filling a similar ecological niche to that of modern day great white sharks, it prowled dark waters, surprising its prey with a rapid strike that overpowered it due to its sheer size, weight, and ferocity. Instead of standardized teeth, it possessed bony teeth-like plates that cut and sheared, very much like giant razor sharp scissors. Its kinetically-powered jaws meant that they could open in less than 1/50th of a second, creating a powerful suction effect that few fish could evade.

Fossil evidence demonstrates that this slow-moving Devonian devil fish was probably quite territorial, battling conspecific rivals for territory and resources. Its plate-like body armor, however, would have helped protect it from violent interactions with both other placoderms and struggling prey.

Possible size comparison between *Dunkleosteus* and man. Copyright © Lochlainn Seabrook.

Dunkleosteus. Copyright © Lochlainn Seabrook.

Dunkleosteus probably died out during the Hangenberg Event, a mass extinction that occurred at the end of the Devonian Period some 358 million years ago. Like most very large prehistoric predators, it may have had difficulty adapting to the many changes then sweeping the earth, as alterations in climate, sea levels, temperature, and oxygen levels modified its environment and resources. Volcanoes and even an asteroid impact may have hastened its demise.

Dunkleosteus terrelli feeding on *Cheirolepis trailli*, a ray-finned fish species from the Devonian period. Copyright © Lochlainn Seabrook.

11
ENTELODON

Entelodon deguilhemi. Copyright © Lochlainn Seabrook.

SCIENTIFIC DETAILS

COMMON NAME: Entelodon.

SCIENTIFIC NAME: *Entelodon deguilhemi*.

ETYMOLOGY: *Entelo* is from the Greek word *enteles*, "complete"; *odon* is Greek for "tooth"; and *deguilhemi* honors French paleontologist Jean-Claude De Guilhem. Full meaning: "Guilhem's complete [set of Entelodon] teeth."

NICKNAME: "Hell pig," "terminator hog."

CLASS: Mammal.

HEIGHT: 4.5 feet to 5 feet at shoulder.

LENGTH: Up to 10 feet.

WEIGHT: Up to 1,200 lbs or more (0.6 tons)

HABITAT: Subtropical to temperate forests, fields, and floodplains.

LIFE SPAN: Possibly up to 25 years.

TIME PERIOD: Late Eocene to early Oligocene, 37 to 28 million years ago.

Head of *Entelodon*. Copyright © Lochlainn Seabrook.

DIETARY CATEGORY: Omnivore, mesopredator, and scavenger.

DIET: Small mammals, reptiles, birds, eggs, carrion, various plant materials such as roots, tubers, and fruit.

HUNTING METHODS: Ambush predation, opportunistic feeder.

TOP SPEED: Possibly 30 mph.

GEOGRAPHIC RANGE: France, Germany, Mongolia, China.

REASONS FOR EXTINCTION: Climate change, habitat loss.

DESCRIPTION

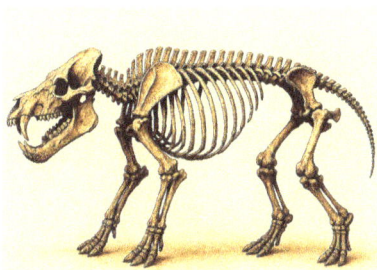

Skeleton of *Entelodon*. Copyright © Lochlainn Seabrook.

The "hell pig," as it is now commonly known, was discovered over 100 years ago in the Quercy Phosphorites Formation located in southwestern France.

This giant cow-sized pig with the body of a pit bull had a mouth full of large, dangerously sharp teeth. This attribute, combined with its impressive robust jaws, allowed it to easily grab and crush live prey. But its muscular 25 inch long skull provided another important survival skill: the ability to effortlessly tear, shred, and process tough vegetable matter.

Though a mesopredator rather than an apex predator, it was truly a force to be reckoned with. While hunting live prey it carefully stalked its victim, waiting in the shadows for an opportunity to pounce. Hurtling itself from the brush at 30 mph, it violently grasped its prey, then quickly pulled it to the ground. Death was quick—though not painless. Fossil toothwear suggests that, as with modern pigs and hogs, scavenging was a more leisurely affair, consisting of devouring whatever edibles were available.

This monstrous 10 foot long artiodactyl is actually more closely related to hippopotamuses, and even whales, than modern pigs. It became extinct during the Oligocene-Miocene Transition approximately 23 million years ago. Climate change brought with it drying and cooling, resulting in the loss of its favorite habitat: open semitropical woodlands.

Size comparison between *Entelodon* and man. Copyright © Lochlainn Seabrook.

Additionally, newly evolving, more modernized mammals, such as true pigs (Suidae) and primitive canids (such as early wolves) began to outcompete *Entelodon*, resulting in a loss of prey species. The end was inevitable.

Entelodon running at top speed: 30 mph, three times faster than the average person can run. Copyright © Lochlainn Seabrook.

Entelodon deguilhemi hunting *Propalaeotherium voigti*, a small horse-like mammal of the Eocene period. Copyright © Lochlainn Seabrook.

12
GIGANTOPITHECUS

Gigantopithecus blackii. Copyright © Lochlainn Seabrook.

SCIENTIFIC DETAILS

COMMON NAME: Giant ape.

SCIENTIFIC NAME: *Gigantopithecus blackii.*

ETYMOLOGY: *Gigantas* is Greek for "giant"; *pithekos* is Greek for "ape"; and *blackii* honors paleoanthropologist Davidson Black. Full meaning: Black's giant ape."

NICKNAMES: "Giganto," "the real King Kong."

CLASS: Mammal.

HEIGHT: Possibly up to 10 feet.

SHOULDER WIDTH: 4 feet or more.

WEIGHT: Up to 1,200 lbs (0.6 tons).

HABITAT: Subtropical to tropical woodlands, karst caves, bamboo forests.

The possible head of *Gigantopithecus blackii.* Copyright © Lochlainn Seabrook.

LIFE SPAN: Possibly up to 50 years.

TIME PERIOD: Pleistocene Epoch, 2 million to 300,000 years ago.

DIETARY CATEGORY: Herbivore, with possibly frugivorous (fruit) and folivorous (leaves) proclivities.

DIET: Bamboo, fruits, leaves, grasses, and hard fibrous seeds, roots, stems, and tubers.

HUNTING METHOD: Herbivorous foraging and scavenging.

TOP SPEED: Perhaps up to 10 mph.

GEOGRAPHIC RANGE: Southeast Asia, including southern China, northern Vietnam, Thailand.

REASONS FOR EXTINCTION: Possible competition with *Homo erectus* and giant pandas, changes in the availability of bamboo.

DESCRIPTION

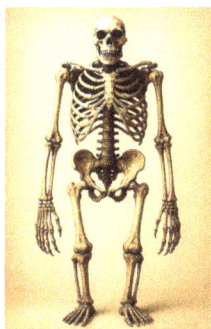

A theoretical illustration of the skeleton of *Gigantopithecus*. Copyright © Lochlainn Seabrook.

Concerning *Homo sapiens* and our long and mysterious evolutionary journey, as a fellow primate and great ape *Gigantopithecus* is the most important creature listed in this book. Unfortunately, we have only a few fossil fragments of this, the world's largest known ape, to study, making any attempt at determining its precise ecology, life history, metrics, and appearance purely hypothetical.

The first trace of *Giganto*, a fossilized molar, was discovered in Hong Kong in a Chinese apothecary shop in 1935 by a German paleontologist. It was being sold as a "dragon tooth." Subsequently this rare find led to the excavation of numerous other teeth, as well as partial mandibles, in caves in southern China.

While describing an animal based on a handful of teeth and incomplete jaws is somewhat hazardous, here is what we know about this extraordinary hominid. Despite its fearsome appearance and size, based on its heavily enameled, large flat teeth, *Gigantopithecus* was almost certainly an herbivore, for such teeth are perfectly adapted for processing and consuming tough fibrous plant material, from stems, roots, and leaves, to fruits, seeds, tubers, and bamboo.

A theoretical size comparison between *Gigantopithecus* and man. Copyright © Lochlainn Seabrook.

Due to its immense weight and height this allegedly "extinct" close cousin of the orangutan was probably a slow-moving vegetarian primate that could transition between walking quadrupedally (fist-walking) and bipedally (on its two hind legs). I use the term "allegedly" because many believe that *Giganto* continues to stalk the earth today as the creature we know as Sasquatch. Indeed, one of its nicknames is the "Asian bigfoot." If true, it would have crossed the Bering Strait land bridge into North America, possibly around 300,000 years ago. As of yet, we have no definitive scientific evidence for this theory. But could *Gigantopithecus*, like the coelacanth, be a Lazarus taxon, one that is not truly extinct, but only appears to be?

Gigantopithecus blackii eating Moso bamboo (*Phyllostachys edulis*) in its native habitat. Could *Giganto* be the legendary Bigfoot, a primate cryptid seen and heard by thousands of credible eyewitnesses around the world, and recorded in the myths of Native Americans for thousands of years? Copyright © Lochlainn Seabrook.

13
GORGONOPS

Gorgonops torvus. Copyright © Lochlainn Seabrook.

SCIENTIFIC DETAILS

COMMON NAME: Gorgonops.

SCIENTIFIC NAME: *Gorgonops torvus*.

ETYMOLOGY: *Gorgon* is the name of a mythological Greek creature; *ops* is Greek for "appearance"; and *torvus* is Latin for "fierce." Full meaning: "Ferocious Gorgon face."

NICKNAME: "Gorgon."

CLASS: Synapsida (an intermediate line that led to mammals).

HEIGHT: 2 feet 6 inches.

LENGTH: Up to 8 feet.

WEIGHT: 132 lbs.

HABITAT: Semi-arid floodplains, marshes.

Head of *Gorgonops*. Copyright © Lochlainn Seabrook.

LIFE SPAN: Possibly up to 20 years.

TIME PERIOD: Late Permian Period, 259 to 252 million years ago.

DIETARY CATEGORY: Carnivore.

DIET: Herbivorous animals, such as dicynodonts and pareiasaurs.

HUNTING METHOD: Solitary ambush predator.

TOP SPEED: 20 mph.

GEOGRAPHIC RANGE: South Africa.

REASONS FOR EXTINCTION: Died out during the Permian-Triassic extinction event some 252 million years ago, which included climate warming, increased volcanic activity, ocean anoxia, and the breakdown of various food chains.

DESCRIPTION

D iscovered in South Africa's Karoo Basin in 1876 by famed British paleontologist Richard Owens, this savage looking solitary apex predator roamed the ancient supercontinent known as Pangaea during the Late Permian Period some 250 million years ago.

Able to outrun most of the smaller animals in its prehistoric domain, it skulked in dense vegetation awaiting an opportunity to pounce on its victim. Darting silently from the brush, it quickly grabbed its prey, puncturing the body with its long saber-like teeth. Powerful unrelenting jaw muscles and slashing teeth spelled certain doom for any animal unlucky enough to come within its grasp.

Skeleton of *Gorgonops*. Copyright © Lochlainn Seabrook.

Neither a dinosaur or a mammal, the iconic dog-like *Gorgon* belonged to the synapsid class, a branch on the evolutionary tree that led to the rise of mammals (including us). Its territorial habitat included wetlands, open forested regions, and fluvial mudflats surrounded by lakes, rivers, and other types of riparian zones.

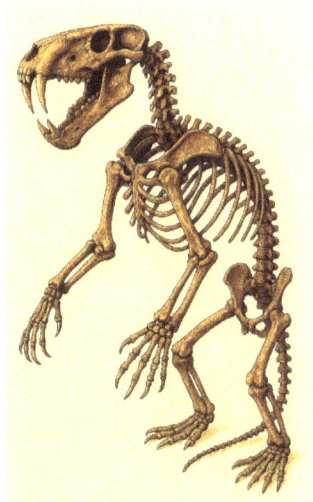

Gorgonops reigned for some 7 million years before succumbing to the devastating effects of the Permian-Triassic extinction event, the most cataclysmal die-off event in our planet's amazing history. Changes in temperature, oceanic oxygen depletion, and the collapse of major

Size comparison between *Gorgonops* and man. Copyright © Lochlainn Seabrook.

ecological niches all contributed to the Gorgon's demise, leaving only its fossilized bones as evidence of its once terrifying existence.

Gorgonops torvus attacking *Diictodon feliceps*. Copyright © Lochlainn Seabrook.

14
HELICOPRION

Helicoprion bessonowi. Copyright © Lochlainn Seabrook.

SCIENTIFIC DETAILS

COMMON NAME: Helicoprion.

SCIENTIFIC NAME: *Helicoprion bessonowi*.

ETYMOLOGY: *Helix* is Greek for "spiral"; *prion* is Greek for "saw"; and *bessonowi* is a Latinized word that seems to derive from the Russian surname Bessonow or Bessonov. Full meaning: "Bessonow's spiral sawed" fish.

NICKNAME: "Buzzsaw shark."

CLASS: Chondrichthyes (cartilaginous fishes).

HEIGHT: About 5 feet at widest part.

LENGTH: Up to 26 feet.

WEIGHT: About 1,110 lbs (0.56 tons).

Head of *Helicoprion*. Copyright © Lochlainn Seabrook.

HABITAT: Shallow temperate to warm epicontinental seas.

LIFE SPAN: Possibly up to 30 years.

TIME PERIOD: Early Permian to early Triassic Periods, about 290 to 250 million years ago.

DIETARY CATEGORY: Carnivore.

DIET: Cephalopods, fish, nautiloids, ammonoids.

HUNTING METHOD: Ambush predator.

TOP SPEED: Slow swimmer, perhaps 5 to 10 mph.

GEOGRAPHIC RANGE: Western USA, Canada, Mexico, Australia, Russia, China, Japan, the Middle East.

REASONS FOR EXTINCTION: Died out during the Permian mass extinction 252 million years ago.

DESCRIPTION

Hypothetical skeleton of *Helicoprion*. Copyright © Lochlainn Seabrook.

First discovered in Russia's Ural Mountains (once part of an ancient sea) in 1889 by Russian geologist Alexander Petrovich Karpinsky, we do not yet have a complete skeleton of this 26 foot long marine goliath, making it difficult to understand its exact body metrics and appearance. However, what we do have has awed and confused scientists and the public for over 100 years: odd fossils of a lower jaw comprised of a whorl-shaped set of teeth that seems to have been capable of quickly rolling up and down, serving as both a predatory grabbing tool and a deadly flesh-slicing implement.

Though based on its theoretical appearance many assume that *Helicoprion* was a prehistoric shark, it was actually a holocephalan, and thus more closely related to modern-day chimaeras (ratfish). Lacking upper jaw teeth and only able to travel at slow to moderate speeds, nonetheless, its enormous size, tremendous physical strength, and living circular "buzzsaw" made it an impressive apex aquatic predator, one capable of ambushing and taking down prey of all sorts, from soft-bodied fish to various species of squid.

A true global citizen, currently *Helicoprion's* fossils have been found in nearly every part of the world except Africa. Having survived for some 22 million years, it is one of the longest living and most widespread prehistoric marine animals on record.

Certainly it must have been incredibly robust: It lived through a number of drastic and violent Permian environmental upheavals, only finally disappearing during the end-of-the-world Permian mass extinction some 250 million years ago—an extinction event so severe that it killed off at least 90 percent of all marine animals. Accompanying, and no doubt hurrying along its demise, was an increase in volcanic activity, global warming, ocean anoxia, and the breakdown of the marine ecosystem and oceanic biodiversity.

Size comparison between *Helicoprion* and man. Copyright © Lochlainn Seabrook.

Helicoprion bessonowi feeding on a *Palaeoniscum freieslebeni*. Copyright © Lochlainn Seabrook.

15
KAPROSUCHUS

Kaprosuchus saharicus. Copyright © Lochlainn Seabrook.

SCIENTIFIC DETAILS

COMMON NAME: Boar crocodile.

SCIENTIFIC NAME: *Kaprosuchus saharicus*.

ETYMOLOGY: *Kapros* is Greek for "boar"; *suchos* is Greek for "crocodile"; and *saharicus* indicates that it was found in the Sahara Desert, which was a tropical region at the time Kaprosuchus lived. Full meaning: "Boar croc of the Sahara."

NICKNAME: "Boar croc."

CLASS: Reptile.

HEIGHT: 4 feet at thickest part.

LENGTH: Up to 20 feet.

WEIGHT: 2,500 lbs (1.25 tons).

Head of *Kaprosuchus*. Copyright © Lochlainn Seabrook.

HABITAT: Tropical marshy lowlands, freshwater rivers, floodplains, and subtropical lake zones.

LIFE SPAN: Possibly up to 50 years.

TIME PERIOD: Late Cretaceous, 95 to 100 million years ago.

DIETARY CATEGORY: Crepuscular carnivore.

DIET: Dinosaurs, fish, turtles, amphibians, other crocodyliforms.

HUNTING METHOD: Aquatic and terrestrial ambush predator.

TOP SPEED: Possibly 25 mph on land, 10 mph in water.

GEOGRAPHIC RANGE: Sahara desert region of Niger.

REASONS FOR EXTINCTION: Climate change, warming ecosystem, competition with other predatory animals such as therapods and crocodyliforms.

DESCRIPTION

Nicknamed the "boar croc" for its massive hog-like tusks, the diet of this mammoth reptilian monstrosity included, among other unlucky creatures, dinosaurs as well as members of its own kind: the crocodyliforms.

Its method of killing was extremely efficient: Whether in water or on land, using its forward-facing, raised 2.5 inch eyes, it silently stalked and ambushed its prey in the dim light during dawn and dusk; extra long legs and a huge muscular tail provided powerful propulsion, allowing the great crocodilian to quickly snatch and firmly hold its quarry. Puncturing the body with its

Skeleton of *Kaprosuchus*. Copyright © Lochlainn Seabrook.

large, interlocking, dagger-like teeth, it tore and lacerated its victim's flesh, gulping down huge strips of meat as it ate. Heavily armored and clocking in at 20 feet in length and over one ton in weight, this prehistoric brute once overwhelmingly dominated the tropical woodlands and waterways of Northern Africa.

Its fossil remains (a single, complete, nearly 25 inch long skull) were first discovered in 2009 in Niger's Sahara Desert by American paleontologist Paul Sereno. Despite the area's dry, hot, desert climate today, at the time *Kaprosuchus* lived there it was a rainy humid region comprised of verdant fern forests and dense vegetation, complete with a diverse array of megafauna and its own shallow sea (the Trans-Saharan seaway). It was in this warm, lush, wet, prey-rich environment that the boar croc thrived, terrorizing its Late Cretaceous neighbors for over 6 million years.

Size comparison between *Kaprosuchus* and man. Copyright © Lochlainn Seabrook.

Not a dinosaur, but sharing a common ancestor with them, like many of the other monsters in this book this terrifying archosaur became extinct due to climate change and the devastating aftereffects that often accompany it: dramatic modifications in temperature along with increasing competition from newly arriving cousins.

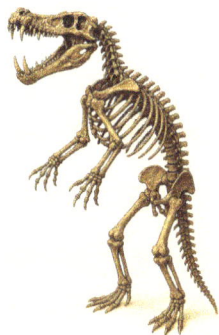

Kaprosuchus saharicus hunting and feeding on *Elrhazosaurus nigeriensis.*. Copyright © Lochlainn Seabrook.

16
KRONOSAURUS

Kronosaurus queenslandicus. Copyright © Lochlainn Seabrook.

SCIENTIFIC DETAILS

COMMON NAME: Kronosaurus.

SCIENTIFIC NAME: *Kronosaurus queenslandicus*.

ETYMOLOGY: *Kronos* derives from the name of the legendary Greek Titan Kronos; *saurus* is from the Greek word *sauros*, meaning "lizard"; and *queenslandicus* is a Latinization of the name of the state of Queensland, Australia, where its fossils were first discovered. Full meaning: "The Kronos lizard from Queensland."

NICKNAME: "Kronos."

CLASS: Reptile.

HEIGHT: 5 feet at thickest section.

LENGTH: Possibly up to 36 feet.

WEIGHT: 28,660 lbs (13 tons).

Head of *Kronosaurus*. Copyright © Lochlainn Seabrook.

HABITAT: Warm shallow inland seas that once covered Australia.

LIFE SPAN: Possibly up to 50 years.

TIME PERIOD: Early Cretaceous, about 125 to 100 million years ago.

DIETARY CATEGORY: Carnivore.

DIET: Fish, cephalopods, turtles, other marine reptiles such as plesiosaurs.

HUNTING METHOD: Sudden and violent ambush attacks.

TOP SPEED: Possibly up to 25 mph.

GEOGRAPHIC RANGE: Australia.

REASONS FOR EXTINCTION: Climate change, competition.

DESCRIPTION

Skeleton of *Kronosaurus*. Copyright © Lochlainn Seabrook.

With a skull length of up to 9 feet, two sets of sharp conical teeth the size of bananas, the ability to launch explosive ambush attacks, and a bite force that was one of the most powerful of any animal during the early Cretaceous, this macabre marine monster could hold its own against nearly any oceangoing creature of its day.

An apex predator belonging to the plesiosaur subgroup known as the pliosaurs, it possessed a short neck, a massive thick head, a streamlined body, four muscular flippers, and keen olfactory organs, all which gave it the flexibility to feed on other large oceanic animals, including fish, squid, ammonites, and one of its smaller cousins, the plesiosaur. Though its teeth lacked the razor-like cutting edges of, for example, modern sharks, its dentition—designed instead for gripping—could easily grab and hold its struggling prey, after which it was swallowed whole.

Kronos' four large flippers gave this obligate aquatic reptile fantastic agility underwater. Yet this extreme maneuverability came at a cost: It could not lay eggs on land like dinosaurs and other oviparous terrestrial animals. Hence, this viviparous beast gave birth to live young in the water.

Discovered in Queensland, Australia, in 1899 by a cattle rancher named Andrew Crombie, *Kronosaurus* survived for an astonishing 25 million years before dying out due to the evolution of newer more specialized

Size comparison between *Kronosaurus* and man. Copyright © Lochlainn Seabrook.

marine reptiles, alterations to its ecosystem, disruption of the food chain, and climactic changes that directly impacted its habitat: temperature, seawater chemistry, and sea levels.

Kronosaurus queenslandicus feeding on *Platypterygius australis.* Copyright © Lochlainn Seabrook.

17
LIVYATAN

Livyatan melvillei. Copyright © Lochlainn Seabrook.

SCIENTIFIC DETAILS

COMMON NAME: Giant killer sperm whale.

SCIENTIFIC NAME: *Livyatan melvillei.*

ETYMOLOGY: *Livyatan* is the Hebrew word for "leviathan," and *melvillei* honors American novelist Herman Melville, author of *Moby Dick*. Full meaning: "Melville's leviathan."

NICKNAME: "The macroraptorial sperm whale."

CLASS: Mammal.

HEIGHT: 12 feet at thickest point.

LENGTH: Up to 60 feet.

WEIGHT: 124,000 lbs (62 tons).

HABITAT: Coastal deep waters and open oceans.

LIFE SPAN: Perhaps 60 years.

Head of *Livyatan*. Copyright © Lochlainn Seabrook.

TIME PERIOD: Miocene Epoch, 9 to 13 million years ago.

DIETARY CATEGORY: Carnivore.

DIET: Seals, large fish, giant squid, other whales.

HUNTING METHOD: Aggressive ambush predation.

TOP SPEED: Possibly 25 mph.

GEOGRAPHIC RANGE: Peru (relatives found in Chile and Argentina).

REASONS FOR EXTINCTION: Climate change, sea current alterations, water temperature cooling, reduction in prey species, competition with the giant shark megalodon.

DESCRIPTION

This 10 million year old sperm whale-like cetacean was among the most frightening aquatic beasts of the Miocene Epoch. Clocking in at over 60 tons, 60 feet in length, and 12 feet in height, and possessing a 10 foot long skull packed with 40

Skeleton of *Livyatan*. Copyright © Lochlainn Seabrook.

teeth that were as long as a laptop computer is wide (and were among the largest biting teeth known), it reigned over its watery South American domain for some 2 to 3 million years.

A fearsome predatory ambush hunter specializing in stealth, power, and speed, it easily caught, killed, and devoured its preferred prey: large fish and giant squid, along with other marine mammals, including seals and even members of its own kind. Its most lethal weapons were its ability to precisely echolocate, its massive head, giving it immense ramming power, and two sets of sharp, conical, 15 inch teeth set in powerful upper and lower jaws. These deadly tools gave it the capacity to quickly find, snatch, hold, tear, and crush the flesh and bone of its unlucky victims.

Named for the monstrous "leviathan" in the Book of Job, this giant prehistoric marine marauder was first discovered in 2008 in the Pisco Formation of southern Peru. Once a land mammal whose vestigial pelvic bones betray its terrestrial origins, it probably occupied a similar ecological niche as its infamous Miocene companion: the world's largest known shark, megalodon, a competitive species that eventually contributed to *Livyatan's* extinction. Other causes involved dramatic climactic downshifts in water temperature, loss of food items, decline of marine productivity, and severe modification of oceanic currents.

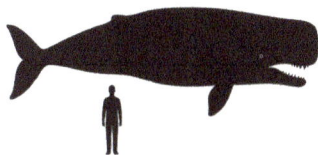

Size comparison between *Livyatan* and man. Copyright © Lochlainn Seabrook.

Besides its frightful visage—one that may have made it look more like a leopard seal than a whale—this terrifying living torpedo possessed one other outstanding trait: Not only did *Livyatan* rival both *Tyrannosaurus rex* and megalodon in size and ferocity, its powerful jaws possessed one of the strongest bite forces on record, easily exceeding 3,600 psi.

Livyatan melvillei feeding on giant squid. Copyright © Lochlainn Seabrook.

18
MAUISAURUS

Mauisaurus haasti. Copyright © Lochlainn Seabrook.

SCIENTIFIC DETAILS

COMMON NAME: Mauisaurus.

SCIENTIFIC NAME: *Mauisaurus haasti*.

ETYMOLOGY: *Maui* is the name of a Maori god; *saurus* is the Greek word for "lizard"'; and *hassti* honors geologist Julius von Haast who discovered the first fossil. Full meaning: "Maui and Haast's Lizard."

NICKNAME: "Loch Ness monster."

CLASS: Reptile.

HEIGHT: Body depth, 5 feet; with outstretched neck, 12 feet.

LENGTH: Up to 27 feet.

WEIGHT: Possibly 2 to 3 tons.

HABITAT: Shallow coastal waters.

Head of *Mauisaurus*. Copyright © Lochlainn Seabrook.

LIFE SPAN: Possibly up to 80 years.

TIME PERIOD: Late Cretaceous, 80 to 69 million years ago.

DIETARY CATEGORY: Carnivorous.

DIET: A piscivore, it fed mainly on fish.

HUNTING METHOD: Stealth and ambush hunter.

TOP SPEED: Possibly 8 mph.

GEOGRAPHIC RANGE: New Zealand.

REASONS FOR EXTINCTION: Climate change, volcanic activity, sea level alterations, and possibly a large asteroid or comet impact 66 million years ago near the Yucatán Peninsula, Mexico.

DESCRIPTION

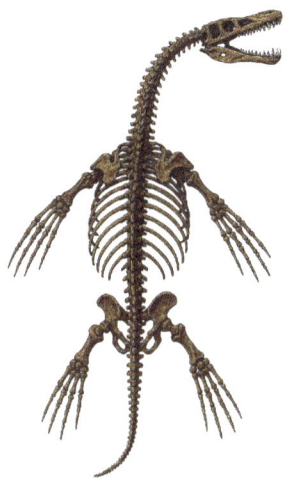

Skeleton of *Mauisaurus*. Copyright © Lochlainn Seabrook.

A classic "sea monster" in every sense of the term, this frightening marine reptile roamed the ancient oceans of the Southern Hemisphere, warm shallow seas that once covered parts of what is now New Zealand. Its large powerful 4 to 5 foot flippers, designed for slow, long distance, open ocean swimming, gave the beast the ability to "fly" underwater, similar to the swimming mode of modern sea turtles.

Due to the enormous drag produced by its tear-shaped body and long neck, Mauisaurus was an efficient ambush predator rather than a high speed chase-and-attack hunter. This fantastical pescivore was incredibly agile and maneuverable thanks to its 66 cervical vertebrae, giving its spine great flexibility. Paddling silently in the dark depths, it violently snatched its prey using two sets of conical 2 inch teeth designed for grasping slippery fish. It then promptly swallowed its victims whole.

Arguably the most recognizable of the elasmosaurid plesiosaurs, Mauisaurus was discovered by geologist Julius von Haast in 1877 near Gore Bay, New Zealand. Though mainstream science maintains that Mauisaurus died out during the massive Cretaceous–Paleogene extinction event 66 million years ago, some believe that it lives on today as the fabled Loch Ness monster. In fact, numerous eyewitness accounts, many from highly persuasive individuals, match this long-necked plesiosaur

Size comparison between *Mauisaurus* and man. Copyright © Lochlainn Seabrook.

perfectly. Unfortunately for believers, however, there is currently no fossil evidence to sustain the theory. The mystery continues!

Left: Is a plesiosaur like Mauisaurus actually the Loch Ness monster? Many people think so. This hypothetical illustration portrays the marine reptile in a Scottish loch. Copyright © Lochlainn Seabrook.

Mauisaurus haasti feeding on *Enchodus petrosus*. Copyright © Lochlainn Seabrook.

19

MEGALODON

Otodus megalodon. Copyright © Lochlainn Seabrook.

SCIENTIFIC DETAILS

COMMON NAME: Megalodon.

SCIENTIFIC NAME: *Otodus megalodon.*

ETYMOLOGY: *Otodus* is from the Greek word *ot* meaning "ear" (shaped); *odous* is Greek for "tooth"; *megas* is Greek for "great." Full meaning: "Ear-shaped great tooth."

NICKNAME: "Megatooth shark."

CLASS: Chondrichthyes (cartilaginous fishes).

HEIGHT: Dorsal fin to belly, 12 feet.

LENGTH: Up to 80 feet.

WEIGHT: Up to 100 tons.

Head of *megalodon.* Copyright © Lochlainn Seabrook.

HABITAT: Warm coastal waters and open-ocean environments.

LIFE SPAN: Up to 70 years.

TIME PERIOD: Early Miocene, 20 million to 3.6 million years ago.

DIETARY CATEGORY: Carnivore.

DIET: Whales, marine mammals, large fish, other sharks.

HUNTING METHOD: Apex ambush predator.

TOP SPEED: Possibly up to 16 mph.

GEOGRAPHIC RANGE: Everywhere except Antarctica.

REASONS FOR EXTINCTION: Climate cooling, changes in sea level, loss of prey populations, competition with smaller more agile sharks.

DESCRIPTION

Skeleton of megalodon. Copyright © Lochlainn Seabrook.

This, the world's biggest shark, ruled our planet's oceans for some 17 million years, a threat to anything and everything that was unfortunate enough to cross its path. A macrophagous predator, it fed only on other large creatures, from sperm whales, sea lions, and dolphins, to baleen whales, walruses, and manatee-like animals.

Lurking in the opaque depths of the sea, it rushed up toward the surface, smashing into its prey with its 80 foot long, 100 ton body. Opening its mouth 10 feet wide, it clamped its 7 inch serrated teeth down on its victim with the strongest bite force known in the animal kingdom, living or extinct: 41,000 psi. (Compare this with the modern great white shark's bite force of 4,000 psi.) When necessary megalodon cut the tails and flippers off its prey in order to disable it, then attacked the chest area, delivering a fatal bite to the vital organs. Shaking its head violently, it tore off huge chunks of the flesh of its quarry, gulping them down in massive mouthfuls the size of a modern compact car.

Though its teeth were known to the ancient Greeks and Romans (who mistook them for petrified "dragon's tongues"), megalodon was not scientifically described until 1843, when Swiss naturalist Louis Agassiz placed it in the same genus as the great white shark, calling it *Carcharodon megalodon*. This classification, however, has since been changed to *Otodus megalodon*, a reflection of newer more accurate analyses.

Truly no other beast in earth's history was as terrifying as the marine monstrosity we know today as megalodon. Yet, according to some, it might still be prowling our oceans. While mainstream science holds that it went extinct approximately 3.6 million years ago due to the disastrous effects of climate change, countless plausible eyewitnesses have come forward over

Size comparison between megalodon and man. Copyright © Lochlainn Seabrook.

the years claiming to have seen it, or something like it, with many sightings having occurred in the Sea of Cortez.

Otodus megalodon feeding on a 16 foot *Herpetocetus transatlanticus*. Copyright © Lochlainn Seabrook.

20

MOSASAURUS

Mosasaurus hoffmannii. Copyright © Lochlainn Seabrook.

SCIENTIFIC DETAILS

COMMON NAME: Mosasaur.

SCIENTIFIC NAME: *Mosasaurus hoffmannii*.

ETYMOLOGY: *Mosa* is Latin for the Meuse River (Netherlands); *sauros* is Greek for "lizard"; and *hoffmannii* honors Dutchman Pieter Hoffmann, the discoverer of the first fossils (late 18th Century). Full meaning: "Hoffmann's lizard of the Meuse River region."

NICKNAME: "Tyrant lizard of the seas."

CLASS: Reptile.

HEIGHT: 8 feet at widest part of body.

LENGTH: 56 feet.

WEIGHT: 35,000 lbs (17.5 tons).

HABITAT: Warm shallow seas.

Head of *Mosasaurus*. Copyright © Lochlainn Seabrook.

LIFE SPAN: Possibly up to 50 years.

TIME PERIOD: Late Cretaceous, 70 to 66 million years ago.

DIETARY CATEGORY: Carnivore.

DIET: Large fish, sea turtles, ammonites, other marine reptiles.

HUNTING METHOD: Ambush predator, long-range pursuit.

TOP SPEED: 30 mph.

GEOGRAPHIC RANGE: Europe, North America, North Africa.

REASONS FOR EXTINCTION: Climate change, ocean acidification, collapse of food chain, temperature changes—all resulting from the Cretaceous-Paleogene extinction event.

DESCRIPTION

O ne of the scariest and most dangerous beasts in this entire book, this prehistoric extinct marine reptile weighed in at a staggering 17.5 tons, grew to 56 feet in length, had a nearly 6 foot long skull, and measured 8 feet tall from back to belly.

One of the primary apex predators of the Late Cretaceous, this powerful streamlined swimmer fed on fellow mosasaurs, along with other large oceanic animals, from large fish to sea turtles. It was an adaptable hunter, able to ambush its quarry or chase it down over long distances at speeds of up to 30 mph, propelled

Skeleton of *Mosasaurus*. Copyright © Lochlainn Seabrook.

along by its muscular shark-like tail. Its snake-like double-hinged jaws and flexible skull bones gave it the ability to swallow large prey whole. Once its bone-crushing 5 inch long conical teeth clamped down on its victim, there was no escape and death came shortly.

Discovered in the Netherlands in the late 1700s by Dutch fossil collector Pieter Hoffmann, mosasaur inhabited the shallow subtropical to tropical seas that covered parts of what is now Europe and North America. This monstrous relative of snakes and lizards thrived for some 4 million years, dominating the mid-water to surface depths that it preferred to haunt, filling an ecological niche similar to modern orcas and great white sharks.

Though not a dinosaur, this gigantic sea-dwelling lizard died out at the same time they did. The cause? Quite probably Mexico's devastating Chicxulub asteroid impact, which occurred during the Cretaceous-Paleogene extinction event around 66 million years ago. Worldwide fires created dust clouds that blocked the sun, interfering with photosynthesis. At least 90 percent of plankton species died off, disrupting the entire marine food

Size comparison between *Mosasaurus* and man. Copyright © Lochlainn Seabrook.

chain from bottom to top. Approximately 75 percent of all animal and plant species disappeared, including *Tyrannosaurus rex*. Another casualty was *Mosasaurus*.

Mosasaurus hoffmannii feeding on an 11 foot long *Plesiosaurus dolichodeirus*. Copyright © Lochlainn Seabrook.

21
QUETZALCOATLUS

Quetzalcoatlus northropi. Copyright © Lochlainn Seabrook.

SCIENTIFIC DETAILS

COMMON NAME: Quetzalcoatlus.

SCIENTIFIC NAME: *Quetzalcoatlus northropi*.

ETYMOLOGY: *Quetzalcoatlus* derives from the name of the Aztec god Quetzalcoatl ("colorful bird-serpent"), and *northropi* honors John K. Northrop, noted American aircraft designer. Full meaning: "Northrop's colorful bird-serpent."

NICKNAMES: "The sky king," "The Texas pterosaur."

CLASS: Reptile.

Head of *Quetzalcoatlus*. Copyright © Lochlainn Seabrook.

HEIGHT: Up to 18 feet.

LENGTH: Up to 27 feet.

WINGSPAN: Up to 40 feet.

WEIGHT: 550 lbs (0.25 tons).

HABITAT: Semiarid plains, coastal floodplains.

LIFE SPAN: Possibly 50 years.

TIME PERIOD: Late Cretaceous, 70 to 66 million years ago.

DIETARY CATEGORY: Carnivore.

DIET: Small vertebrates, juvenile dinosaurs, reptiles, fish.

HUNTING METHOD: Foraging, stabbing and gulping prey.

TOP SPEED: Up to 80 mph.

GEOGRAPHIC RANGE: Southwestern U.S., chiefly Texas.

REASON FOR EXTINCTION: Cretaceous-Paleogene extinction event 66 million years ago.

DESCRIPTION

Skeleton of *Quetzalcoatlus*. Copyright © Lochlainn Seabrook.

This enormous aerial reptile, the largest known flying animal in our planet's history, was first discovered by a University of Texas (Austin) geology graduate student named Douglas A. Lawson. The year was 1971, the location was the Maastrichtian-age Javelina Formation, now part of Big Bend National Park, Texas.

Quetzalcoatlus well-earned its nickname the "sky king": It stood nearly 20 feet tall (the height of a giraffe), was almost 30 feet long (the length of a telephone pole), weighed a quarter of a ton (the weight of a grand piano), and possessed a wingspan of some 40 feet—the width of a basketball court, the length of a standard swimming pool, or the length of an American school bus.

This magnificent stork-like meat-eater ruled both the airspace and the great plains of the southwestern U.S. for some 4 million years, a deathly menace to all small to medium-sized creatures, from juvenile dinosaurs and amphibians, to fish and reptiles. Perfectly at home walking on all fours, when foraging along the ground with its long toothless beak was unsuccessful, it probably turned to scavenging, feasting on carrion left by other predators.

The bone structure of this dynamic soaring reptile was a biomechanical wonder, combining strength, lightness, durability, and flexibility, all which enabled it to reach speeds of up to 80 mph while in thermal gliding mode. One might well wonder how this massive flying reptile got airborne. While standing on all four limbs it used a method known as the quadrupedal launch: a burst-like spring motion that catapulted it into the air, after which it snapped open its flapping wings, soaring skyward.

Sadly, we will never know the thrill of watching the awe-inspiring Quetzalcoatlus roam our skies, for it died out during the Cretaceous-Paleogene extinction event some 66 million years ago.

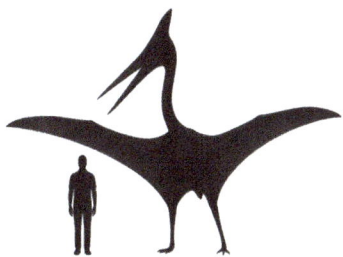

Size comparison between *Quetzalcoatlus* and man. Copyright © Lochlainn Seabrook.

Quetzalcoatlus northropi feeding on *Saurornitholestes langstoni*. Copyright © Lochlainn Seabrook.

22
SMILODON

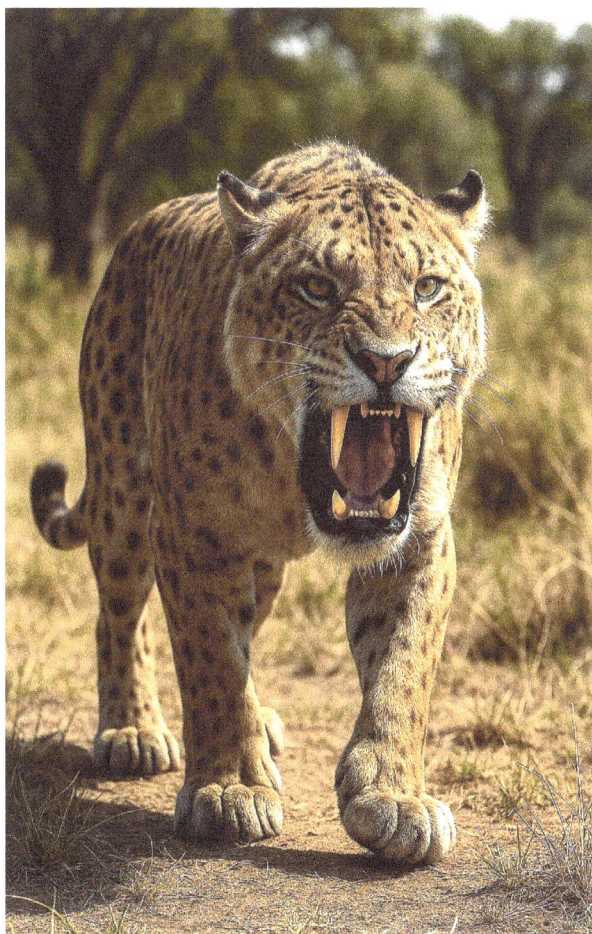

Smilodon fatalis. Copyright © Lochlainn Seabrook.

SCIENTIFIC DETAILS

COMMON NAME: Saber-toothed cat.

SCIENTIFIC NAME: *Smilodon fatalis.*

ETYMOLOGY: *Smil* is from the Greek word *smile* meaning "carving knife"; *odon* is Greek for "tooth"; and *fatalis* is the Latin word for "deadly." Full meaning: "Deadly knife tooth."

NICKNAME: "Sabertooth."

CLASS: *Felidae* (cats).

HEIGHT: 3 feet 6 inches.

LENGTH: Up to 7 feet from nose to tail tip. (Tail length was only 12 inches.)

WEIGHT: Up to 620 lbs (for males).

HABITAT: Plains, open woodlands, grassland, scrubland, valleys, brushland.

Head of *Smilodon*. Copyright © Lochlainn Seabrook.

LIFE SPAN: Probably 12 years in the wild. (Might have lived up to 20 years in captivity.)

TIME PERIOD: Pleistocene, 2.5 million to 10,000 years ago.

DIETARY CATEGORY: Obligate carnivore.

DIET: Horses, mammoths, mastodons, bison, camels.

HUNTING METHOD: Ambush predator.

TOP SPEED: Up to 30 mph.

GEOGRAPHIC RANGE: North and South America.

REASONS FOR EXTINCTION: End of Ice Age, climate change.

DESCRIPTION

O ften mistakenly called a "saber-toothed tiger," *Smilodon* is not closely related to modern tigers (*Panthera tigris*), making the name scientifically inaccurate. Though the two species descend from a common feline-like ancestor and are thus indeed distant cousins, their families (tigers—from the subfamily Pantherinae, and *Smilodon*—from the subfamily Machairodontinae) diverged some 20 to 23 million years ago. Its closest living relative appears to be the American mountain lion (*Puma concolor*), more commonly known as a cougar.

Smilodon skeleton. Copyright © Lochlainn Seabrook.

While the long-tailed modern mountain lion is flexible and lithe with a streamlined low-slung profile, the iconic saber-toothed cat was short-tailed and stocky with massive front quarters—a body design built, not for speed and agility, but for explosive power and shock-and-awe ambush predation. Its enormous somewhat fragile canines, for which it was named, were not used for tearing and shredding, but rather for stabbing and puncturing. After sneaking up and pouncing on its prey, it grabbed and pinned it with its muscular forelimbs, using its robust jaws to sink its 7 inch (11 inches long including the root) into its victim's neck or abdominal region. Prey items included bison, horses, mammoths, and camels.

Since 1842, when the first *Smilodon* fossil was found by Danish paleontologist Peter W. Lund in Brazil, many thousands of saber-

Size comparison between *Smilodon* and man. Copyright © Lochlainn Seabrook.

toothed cat remains have been discovered, particularly at California's renowned La Brea Tar Pits. This fact, along with its famous ferocious fangs, has made it one of the most recognizable and popular extinct prehistoric animals in the world.

While we may shudder at the thought of living alongside *Smilodon*, many of our human ancestors did just that; among them the Paleo-Indians, pre-Clovis people, and Clovis people of North and South America (lived some 13,000 to 20,000 years ago). Yet, the end of the Ice Age spelled the end of *Smilodon* as well, with the inevitable parallel aftereffects: climate change, loss of prey, and competition with both humans and other apex predators.

Smilodon fatalis attacking a nearly 4 ton *Bison antiquus*. Copyright © Lochlainn Seabrook.

23
SPINOSAURUS

Spinosaurus aegyptiacus. Copyright © Lochlainn Seabrook.

SCIENTIFIC DETAILS

COMMON NAME: Spinosaurus.

SCIENTIFIC NAME: *Spinosaurus aegyptiacus*.

ETYMOLOGY: *Spino* derives from the Latin word *spina*, "spine"; *saurus* is from the Greek word *sauros*, "lizard"; and *aegyptiacus* is a Latinized form of Egypt, where it was first found. Full meaning: "spine lizard from Egypt."

NICKNAMES: "The sail-backed lizard," "spine lizard."

CLASS: Reptile.

HEIGHT: From sail tip down, 20 feet; from hips down, 10 feet.

LENGTH: Up to 52 feet.

WEIGHT: Up to 20,000 lbs (10 tons).

HABITAT: Wet tropical floodplains, deltas, riverine regions.

LIFE SPAN: Possibly up to 30 years.

Head of *Spinosaurus*. Copyright © Lochlainn Seabrook.

TIME PERIOD: Late Cretaceous, 99 to 93.5 million years ago.

DIETARY CATEGORIES: Riparian carnivore and piscivore.

DIET: Large fish, small to mid-sized dinosaurs and pterosaurs.

HUNTING METHODS: Stalk and pounce, scavenging.

TOP SPEED: Possibly 10 mph.

GEOGRAPHIC RANGE: Northern Africa.

REASONS FOR EXTINCTION: The Cretaceous–Paleogene mass extinction event around 66 million years ago.

DESCRIPTION

Skeleton of *Spinosaurus*. Copyright © Lochlainn Seabrook.

The discovery of the first Spinosaurus fossils in 1912 in the Bahariya Oasis of the Western Desert, Egypt, was heralded as a great advancement in our knowledge of dinosaurs. However, these, the original remains, were lost in 1944 at the time of the Allied bombing of Munich, Germany, during World War II. Fortunately, many new fossils of this frightening but important dinosaur have been discovered since then.

The first known semi-aquatic therapod, one that highlights the astounding diversity of prehistoric animals in North Africa, this gigantic, 52 foot long, piscivorous predator evolved to live near water, probably hunting large fish in the shallows of lakes, rivers, and estuaries. When fishing was poor it is likely to have hunted small dinosaurs, resorting to scavenging when required. Its narrow crocodilian-like snout, which contained movement-sensing cells, was filled with conical non-serrated teeth, well designed for capturing and holding slime-covered aquatic prey. Able to walk both bipedally and quadrupedally, it probably favored the former while stalking swampy shallows. Mid-water shoreline hunting would have been expedited by its vertically flat, newt-like tail, which aided in propulsion and balance. Large forelimbs with deeply hooked claws were useful for grappling with slippery struggling prey.

A native of what is now Egypt, Morocco, and Algeria (and possibly Tunisia and Niger), its giant sail-like neural back spines probably evolved to store energy, regulate body temperature, or for sexual display—or perhaps a combination of all three.

Spinosaurus survived for some 6 million years (in contrast we modern humans have only existed for about 300,000 years), succumbing to the world-altering Cretaceous–Paleogene mass extinction event around 66 million years ago. Mexico's Chicxulub asteroid impact triggered global firestorms, nuclear-winter conditions, acid rain, and, of course, disastrous climate change. When it was all over, Spinosaurus (along with 100 percent of all other non-avian dinosaurs) was gone, a calamitous casualty of Nature's mindless whim.

Size comparison between *Spinosaurus* and man. Copyright © Lochlainn Seabrook.

Spinosaurus aegyptiacus feeding on *Ouranosaurus nigeriensis*. Copyright © Lochlainn Seabrook.

24

TITANOBOA

Titanoboa cerrejonensis. Copyright © Lochlainn Seabrook.

SCIENTIFIC DETAILS

COMMON NAME: Titanoboa.

SCIENTIFIC NAME: *Titanoboa cerrejonensis.*

ETYMOLOGY: *Titan* derives from the Greek word *titan*, "giant"; *boa* is Latin for "large snake"; *cerrejon* is the name of a geological formation in Columbia; and the suffix *ensis* is Latin for "from." Full meaning: "Giant snake from Cerrejon."

NICKNAME: "Titanic boa."

CLASS: Reptile.

BODY DIAMETER: Over 3 feet.

Head of *Titanoboa*. Copyright © Lochlainn Seabrook.

LENGTH: Nearly 50 feet.

WEIGHT: 2,500 lbs (1.25 tons).

HABITAT: Hot, humid, dense, neotropical swamps, rain forests.

LIFE SPAN: Possibly up to 30 years.

TIME PERIOD: Middle to Late Paleocene, about 60 to 58 million years ago.

DIETARY CATEGORY: Carnivore.

DIET: Fish, crocodilian relatives.

HUNTING METHOD: Semi-aquatic ambush predator.

TOP SPEED: Up to 5 mph on land, up to 15 mph in water.

GEOGRAPHIC RANGE: Modern day northern Columbia.

REASONS FOR EXTINCTION: Climate change, cooling temps.

DESCRIPTION

The first fossil remains of this terrifying serpent were discovered in 2002 in Columbia's Cerrejón coal mines by joint teams led by Carlos Jaramillo and Jonathan Bloch. The largest snake to have ever lived, at some 50 feet in length it was twice as long as our longest modern snake, the 25 foot long reticulated python (*Malayopython reticulatus*)—absolutely dwarfing it, not only in length but also in both weight and girth.

Skeleton of *Titanoboa*. Copyright © Lochlainn Seabrook.

Fortunately for us humans, we did not live during the Middle to Late Paleocene Epoch, for without question *Homo sapiens* would have been on Titanoboa's menu: As the shocking size comparison below illustrates, a snake of its dimensions could easily ambush, encircle, constrict, crush, suffocate, and swallow whole any human alive today.

A semi-aquatic apex predator that inhabited thick, wet, humid, swampy regions, studies of its skull reveal a propensity toward piscivory. However, its diet no doubt also included crocodiles, turtles, birds, and mammals, or whatever else it could catch when its freshwater ichthyic quarry was scarce or unavailable. As giant sharks, such as megalodon, would not appear for another 37 million years, it is likely that this ferocious alpha-level carnivore filled a similar ecological niche during the Paleocene.

It reigned over its tropical domain for some 4 million years. Around 56 million years ago, however, global climate began to cool, greatly affecting ectothermic (cold-blooded) animals like Titanoboa, who could not thermoregulate, but instead relied on the external environment to help stabilize body temperature. Alterations in temperature created simultaneous changes in habitat, prey diversity, water levels, and oxygenation, as well as an increase in competitive species—all spelling certain doom for the "titanic boa constrictor" from Columbia.

Size comparison between *Titanoboa* and man. Copyright © Lochlainn Seabrook.

Titanoboa cerrejonensis hunting an 8 foot long *Carodnia vieirai*. Copyright © Lochlainn Seabrook.

25
TYRANNOSAURUS

Tyrannosaurus rex. Copyright © Lochlainn Seabrook.

SCIENTIFIC DETAILS

COMMON NAME: Tyrannosaurus.

SCIENTIFIC NAME: *Tyrannosaurus rex.*

ETYMOLOGY: *Tyranno* is from the Greek word *tyrannos*, "tyrant"; *saurus* is Greek for "lizard"; and *rex* is Latin for "king." Full meaning: "Tyrant lizard king."

NICKNAMES: "T. rex," "king of the dinosaurs."

CLASS: Reptile.

HEIGHT: Up to 20 feet.

LENGTH: Possibly up to and over 42 feet.

WEIGHT: Up to 20,000 lbs (10 tons).

Head of *Tyrannosaurus rex.*
Copyright © Lochlainn Seabrook.

HABITAT: Coastal plains, river valleys, subtropical forests.

LIFE SPAN: Possibly up to 30 years.

TIME PERIOD: Late Cretaceous, some 68 to 66 million years ago.

DIETARY CATEGORY: Obligate carnivore.

DIET: Mainly herbivorous dinosaurs.

HUNTING METHOD: Apex predation, opportunistic scavenging.

TOP SPEED: Perhaps 15-20 mph.

Skeleton of *Tyrannosaurus rex.*
Copyright © Lochlainn Seabrook.

GEOGRAPHIC RANGE: Western North America (Wyoming, North Dakota, Utah, Colorado, Texas, New Mexico, Montana).

REASON FOR EXTINCTION: The Cretaceous-Paleogene extinction event.

DESCRIPTION

E veryone's favorite dinosaur was first discovered in 1902 in Montana's Hell Creek Formation by Barnum Brown, then the assistant curator at the American Museum of Natural History. Arguably the most frightening animal to have ever lived, *T. rex*, as it is often light-heartedly called, was a massive beast: some 42 feet in length, 20 feet in height, and a staggering 10 tons in weight. Its giant skull alone was over 5 feet long, overt evidence of extra powerful jaw muscles capable of delivering one of the strongest bite forces known in nature: over 12,000 psi.

Its front legs were small, only 3 feet long. However, this lack was more than compensated for by a mouth full of some 60 banana-shaped, deeply rooted 12 inch teeth; serrated pain-inflicting daggers perfectly designed for grasping, puncturing flesh, slicing through tough hides, and crushing hard bones. An obligate carnivorous apex predator ruling the top of the food chain, its diet was comprised of vegetarian dinosaurs, creatures such as *Triceratops horridus*, *Thescelosaurus neglectus*, *Ankylosaurus magniventris*, *Edmontosaurus annectens*, *Pachycephalosaurus wyomingensis*, *Ornithomimus velox*, and even other *T. rexes*.

While adults probably hunted alone, juveniles may have formed social hunting bands in order to better ensure survival. Forward-facing eyes, which gave it excellent depth perception and stereoscopic vision, along with perhaps the best sense of smell of any animal in history, gave it an enormous advantage over its prey. With such a deadly biological arsenal at its disposal we can be sure that the hunting success rate of this nightmarish monster was close to 100 percent.

Size comparison between *Tyrannosaurus rex* and man. Copyright © Lochlainn Seabrook.

After terrorizing its Late Cretaceous neighbors for some 2 million years—leaving little but carnage in its wake—the supreme reign of this veritable "king of the dinosaurs" was finally overthrown by a small cosmic "pebble" from outer space, an asteroid, making it yet another victim of the world-altering Cretaceous-Paleogene extinction event that wiped out 100 percent of all non-avian dinosaurs. The demise of *Tyrannosaurus rex* and its kind, however, allowed the evolution of mammals, and by association, human beings, to flourish into the present day.

Tyrannosaurus rex hunting a 30 foot long *Triceratops horridus*. Copyright © Lochlainn Seabrook.

TEN BONUS ANIMALS

AMERICAN LION

Lochlainn Seabrook

13 feet long, 5.5 feet tall, 1,050 lbs. Copyright © Lochlainn Seabrook.

BRONTOSCORPIO

3.3 feet long, 10 inches tall, 4 lbs. Copyright © Lochlainn Seabrook.

DAKOSAURUS

16 feet long, 4 feet tall, 2,600 lbs (1.3 tons). Copyright © Lochlainn Seabrook.

INOSTRANCEVIA

13 feet long, 4.5 feet tall, weight 1,300 lbs. Copyright © Lochlainn Seabrook.

LIOPLEURODON

33 feet long, 5 feet tall, 6,000 lbs (3 tons). Copyright © Lochlainn Seabrook.

MEGALANIA

26 feet long, 5 feet tall, 2,000 lbs (1 ton). Copyright © Lochlainn Seabrook.

PHORUSRHACOS

Lochlainn Seabrook

10 feet long, 8 feet tall, 320 lbs. Copyright © Lochlainn Seabrook.

SCUTOSAURUS

8.2 feet long, 4.6 feet tall, weight 2,200 lbs (1.1 tons). Copyright © Lochlainn Seabrook.

TANYSTROPHEUS

Lochlainn Seabrook

20 feet long, 3 feet tall at shoulders, 100 lbs. Copyright © Lochlainn Seabrook.

THALATTOARCHON

30 feet long, 5 feet tall, 8,000 lbs (4 tons). Copyright © Lochlainn Seabrook.

BIBLIOGRAPHY

And Suggested Reading

Tyrannotitan chubutensis. Copyright © Lochlainn Seabrook.

Allen, Emory Adams. *The Prehistoric World: Or, Vanished Races.* Nashville, TN: Central Publishing House, 1885.

Álvarez, Walter. *T. Rex and the Crater of Doom.* New York: Princeton University Press, 1997.

Artis, Edmund Tyrell. *Antediluvian Phytology, Illustrated by a Collection of the Fossil Remains of Plants Peculiar to the Coal Formations of Great Britain.* London, UK: Rodwell and Martin, 1825–1827.

Bakker, Robert T. *Raptor Red.* New York: Bantam Books, 1995.

Benton, Michael J. *When Life Nearly Died: The Greatest Mass Extinction of All Time.* London, UK: Thames and Hudson, 2003.

Brusatte, Stephen L. *Dinosaur Paleobiology.* Oxford, UK: John Wiley and Sons, 2012.

——. *The Rise and Reign of the Mammals: A New History, From the Shadow of the Dinosaurs to Us.* New York: Mariner Books, 2022.

Carroll, Robert L. *Vertebrate Paleontology and Evolution.* New York: W. H. Freeman and Company, 1988.

Cartron, Jean-Luc E., and Jennifer K. Frey (eds.). *Wild Carnivores of New Mexico.* Albuquerque, NM: University of New Mexico Press, 2023.

Chambers, Robert. *Vestiges of the Natural History of Creation.* London, UK: John Churchill, 1844.

Chevalier, Tracy. *Remarkable Creatures.* New York: HarperCollins, 2009.

Chiarelli, Sam. *Dig: A Personal Prehistoric Journey.* Pittston, PA: Hippocampus, 2018.

Clutton-Brock, Juliet (ed.) *The Walking Larder: Patterns of Domestication, Pastoralism, and Predation.* London, UK: Routledge, 1989.

Crichton, Michael. *Dragon Teeth.* New York: HarperCollins, 2017.

Cuvier, Georges. *Recherches sur les ossemens fossiles, où l'on rétablit les caractères de plusieurs espèces d'animaux que les révolutions du globe paraissent avoir détruites.* Paris, France: n.p., 1821–1823.

De Camp, L. Sprague, and Catherine Crook de Camp. *The Day of the Dinosaur.* Garden City, NY: Doubleday, 1968.

Dixon, Dougal. *The New Dinosaurs: An Alternative Evolution.* New York: Salem House, 1988.

Everhart, Michael J. *Oceans of Kansas: A Natural History of the Western Interior Sea.* Bloomington, IN: Indiana University Press, 2017.

Fariña, Richard A., Sergio F. Vizcaíno, and Gerry De Iuliis. *Megafauna: Giant Beasts of Pleistocene South America.* Bloomington, IN: Indiana University Press, 2013.

Fortey, Richard. *Trilobite: Eyewitness to Evolution.* New York: W. W. Norton, 2000.

Frey, Robert C. *Middle and Upper Ordovician Nautiloid Cephalopods of the Cincinnati Arch Region of Kentucky, Indiana, and Ohio.* Washington, D.C.: U.S. Government Printing Office, 1995.

Gittleman, John L. (ed.). *Carnivore Behavior, Ecology, and Evolution.* Ithaca, NY: Cornell University Press, 1996.

Gould, Stephen Jay. *Wonderful Life: The Burgess Shale and the Nature of History.* New York: W. W. Norton, 1989.

Grady, Wayne. *The Bone Museum: Travels into Lost Worlds of Dinosaurs, Mammoths, and Early Humans.* Toronto, Canada: Key Porter Books, 1998.

Grigg, Gordon, and David Kirshner. *Biology and Evolution of Crocodylians.* Ithaca, NY: Cornell University Press, 2015.

Harrington, C. R. (ed.). *Annotated Bibliography of Quaternary Vertebrates of Northern North America.* Toronto, Canada: University of Toronto Press, 2003.

Hone, David W. E. *The Tyrannosaur Chronicles: The Biology of the Tyrant Dinosaurs.* Bloomington, IN: Indiana University Press, 2016.

Horner, John R. *Digging Dinosaurs: The Search That Unraveled the*

Mystery of Baby Dinosaurs. Berkeley, CA: University of California Press, 2009.

Joseph, Frank (ed.). *Discovering the Mysteries of Ancient America: Lost History and Legends, Unearthed and Explored*. Franklin Lakes, NJ: Career Press, 2006.

Kristensen, Hans A., and William G. Bradley. *Dinosaur Ghosts: A True Story of Prehistoric Discovery*. Boston, MA: Little, Brown, 1986.

Kudlinski, Kathleen. *Boy, Were We Wrong About Dinosaurs*. New York: Dial Books, 2014.

Kurtén, Björn. *Before the Indians*. New York: Columbia University Press, 1988.

Lambert, David. *The Field Guide to Prehistoric Life*. New York: Facts on File, 1986.

Larson, Peter, and Kristin Donnan. *Rex Appeal*. Montpelier, VT: Invisible Cities Press, 2002.

Larson, Peter, and Kenneth Carpenter. *Tyrannosaurus Rex, the Tyrant King*. Bloomington, IN: Indiana University Press, 2008.

Lomax, Dean R. *Locked in Time: Animal Behavior Unearthed in 50 Extraordinary Fossils*.

Los Angeles County Museum. *Guide to the Exhibit of Fossil Animals from Rancho La Brea*. Los Angeles, CA: Exposition Park, 1915.

Lull, Richard Swann. *Handbook of Paleontological Techniques: Part I*. New York: McGraw-Hill, 1953.

Mayor, Adrienne. *The First Fossil Hunters: Paleontology in Greek and Roman Times*. Princeton, NJ: Princeton University Press, 2000.

McGowan, Christopher. *Dinosaurs, Spitfires, and Sea Dragons: The Remarkable David Norman*. New York: Pegasus Books, 2014.

Miller, Hugh. *The Cruise of the Betsey: Or, A Summer Ramble Among the Fossiliferous Deposits of the Hebrides*. Edinburgh, Scotland: Johnstone and Hunter, 1857.

——. *The Testimony of the Rocks; Or, Geology in Its Bearings on the Two Theologies, Natural and Revealed*. Edinburgh, Scotland: Johnstone and Hunter, 1857.

Mills, Jim. *Discovering Dinosaurs*. London, UK: Frances Lincoln, 2006.

Naish, Darren. *Walking with Dinosaurs: The Evidence*. London, UK: BBC Books, 1999.

Naish, Darren, and Paul Barrett. *Dinosaurs: How They Lived and Evolved*. London, UK: Natural History Museum, 2016.

Novas, Fernando E. *Dinosaurs of Patagonia*. Bloomington, IN: Indiana University Press, 2009.

Ósmólska, Halszka, Peter Dodson, and David B. Weishampel

(eds.). *The Dinosauria*. Berkeley, CA: University of California Press, 2004.

Owen, Richard. *Memoir on the Vertebrate Fossils of the Secondary Formations of the British Islands*. London, UK: Richard and John E. Taylor, 1846.

——. *History of British Fossil Reptiles* (4 vols.). London, UK: n. p. 1849–1884.

Padian, Kevin. *The Dinosauria: A Natural History*. New York: McGraw-Hill, 1997.

Paul, Gregory S. *Predatory Dinosaurs of the World: A Complete Illustrated Guide*. New York: Simon and Schuster, 1988.

Persons, W. Scott. *Mega Rex: A Tyrannosaurus Named Scotty*. Harbour Publishing, 2024.

Phillips, John. *Figures and Descriptions Illustrative of British Organic Remains*. London, UK: Longman, Rees, Orme, Brown, Green and Longman, 1841–1844.

Prothero, Donald R. *Bringing Fossils to Life: An Introduction to Paleobiology*. New York: Columbia University Press, 2007.

——. *Vertebrate Evolution: From Origins to Dinosaurs and Beyond*. Boca Raton, FL: CRC Press, 2022.

Rea, Tom. *Bone Wars: The Excavation and Celebrity of Andrew Carnegie's Dinosaur*. Pittsburgh, PA: University of Pittsburgh Press, 2001.

Rogers, Raymond R. *Geological Aspects of Deep-Time Paleontology*. Ithaca, NY: Cornell University Press, 2011.

Ruedemann, Rudolf. *Cephalopoda of the Beekmantown and Chazy Formations of the Champlain Basin*. Albany, NY: New York State Education Dept., 1906.

Russell, Dale A. *A Pictorial Guide to the Living Dinosaurs*. Burlington, Ontario, Canada: Queens College Press, 1985.

Sampson, Scott D., and James A. Gauthier (eds.). *Anatomy, Phylogeny, and Palaeobiology of Dinosaurs*. Berkeley, CA: University of California Press, 2000.

Schwimmer, David R. *King of the Crocodylians: The Paleobiology of Deinosuchus*. Bloomington, IN: Indiana University Press, 2002.

Scott, William Berryman. *Some Memories of a Paleontologist*. Princeton, NJ: Princeton University Press, 1939.

Seabrook, Lochlainn. *Aphrodite's Trade: The Hidden History of Prostitution Unveiled*. 1994. Franklin, TN: Sea Raven Press, 2011 ed.

——. *The Goddess Dictionary of Words and Phrases: Introducing a New Core Vocabulary for the Women's Spirituality Movement*. 1997. Franklin, TN: Sea Raven Press, 2010 ed.

——. *Britannia Rules: Goddess-Worship in Ancient Anglo-Celtic Society - An Academic Look at the United Kingdom's Matricentric Spiritual Past.* 1999. Franklin, TN: Sea Raven Press, 2010 ed.

——. *The Book of Kelle: An Introduction to Goddess-Worship and the Great Celtic Mother-Goddess Kelle, Original Blessed Lady of Ireland.* 1999. Franklin, TN: Sea Raven Press, 2010 ed.

——. *Carnton Plantation Ghost Stories: True Tales of the Unexplained from Tennessee's Most Haunted Civil War House!* 2005. Franklin, TN, 2016 ed.

——. *Nathan Bedford Forrest: Southern Hero, American Patriot.* 2007. Franklin, TN, 2010 ed.

——. *Abraham Lincoln: The Southern View.* 2007. Franklin, TN: Sea Raven Press, 2013 ed.

——. *The McGavocks of Carnton Plantation: A Southern History - Celebrating One of Dixie's Most Noble Confederate Families and Their Tennessee Home.* 2008. Franklin, TN, 2011 ed.

——. *Christmas Before Christianity: How the Birthday of the "Sun" Became the Birthday of the "Son."* Franklin, TN: Sea Raven Press, 2010.

——. *A Rebel Born: A Defense of Nathan Bedford Forrest.* 2010. Franklin, TN: Sea Raven Press, 2011 ed.

——. *Everything You Were Taught About the Civil War is Wrong, Ask a Southerner!* 2010. Franklin, TN: Sea Raven Press, 2024 ed.

——. *The Quotable Jefferson Davis: Selections From the Writings and Speeches of the Confederacy's First President.* Franklin, TN: Sea Raven Press, 2011.

——. *The Quotable Robert E. Lee: Selections From the Writings and Speeches of the South's Most Beloved Civil War General.* Franklin, TN: Sea Raven Press, 2011 Sesquicentennial Civil War Edition.

——. *Lincolnology: The Real Abraham Lincoln Revealed In His Own Words.* Franklin, TN: Sea Raven Press, 2011.

——. *The Unquotable Abraham Lincoln: The President's Quotes They Don't Want You To Know!* Franklin, TN: Sea Raven Press, 2011.

——. *Honest Jeff and Dishonest Abe: A Southern Children's Guide to the Civil War.* Franklin, TN: Sea Raven Press, 2012.

——. *Encyclopedia of the Battle of Franklin - A Comprehensive Guide to the Conflict that Changed the Civil War.* Franklin, TN: Sea Raven Press, 2012.

——. *The Quotable Nathan Bedford Forrest: Selections From the Writings and Speeches of the Confederacy's Most Brilliant Cavalryman.* Spring Hill, TN: Sea Raven Press, 2012.

——. *Forrest! 99 Reasons to Love Nathan Bedford Forrest.* Spring Hill, TN: Sea Raven Press, 2012.

——. *Give 'Em Hell Boys! The Complete Military Correspondence of*

Nathan Bedford Forrest. Spring Hill, TN: Sea Raven Press, 2012.

——. *The Constitution of the Confederate States of America Explained: A Clause-by-Clause Study of the South's Magna Carta.* Spring Hill, TN: Sea Raven Press, 2012 Sesquicentennial Civil War Edition.

——. *The Great Impersonator: 99 Reasons to Dislike Abraham Lincoln.* Spring Hill, TN: Sea Raven Press, 2012.

——. *The Old Rebel: Robert E. Lee As He Was Seen By His Contemporaries.* Spring Hill, TN: Sea Raven Press, 2012 Sesquicentennial Civil War Edition.

——. *The Quotable Stonewall Jackson: Selections From the Writings and Speeches of the South's Most Famous General.* Spring Hill, TN: Sea Raven Press, 2012 Sesquicentennial Civil War Edition.

——. *Saddle, Sword, and Gun: A Biography of Nathan Bedford Forrest for Teens.* Spring Hill, TN: Sea Raven Press, 2013.

——. *Jesus and the Law of Attraction: The Bible-Based Guide to Creating Perfect Health, Wealth, and Happiness Following Christ's Simple Formula.* Franklin, TN: Sea Raven Press, 2013.

——. *The Bible and the Law of Attraction: 99 Teachings of Jesus, the Apostles, and the Prophets.* Franklin, TN: Sea Raven Press, 2013.

——. *The Alexander H. Stephens Reader: Excerpts From the Works of a Confederate Founding Father.* Spring Hill, TN: Sea Raven Press, 2013.

——. *The Quotable Alexander H. Stephens: Selections From the Writings and Speeches of the Confederacy's First Vice President.* Spring Hill, TN: Sea Raven Press, 2013 Sesquicentennial Civil War Edition.

——. *Christ Is All and In All: Rediscovering Your Divine Nature and the Kingdom Within.* Franklin, TN: Sea Raven Press, 2014.

——. *Jesus and the Gospel of Q: Christ's Pre-Christian Teachings as Recorded in the New Testament.* Franklin, TN: Sea Raven Press, 2014.

——. *Give This Book to a Yankee! A Southern Guide to the Civil War for Northerners.* Spring Hill, TN: Sea Raven Press, 2014.

——. *The Articles of Confederation Explained: A Clause-by-Clause Study of America's First Constitution.* Spring Hill, TN: Sea Raven Press, 2014.

——. *Confederate Blood and Treasure: An Interview With Lochlainn Seabrook.* Spring Hill, TN: Sea Raven Press, 2015.

——. *Nathan Bedford Forrest and the Battle of Fort Pillow: Yankee Myth, Confederate Fact.* Spring Hill, TN: Sea Raven Press, 2015.

——. *Everything You Were Taught About American Slavery War is Wrong, Ask a Southerner!* Spring Hill, TN: Sea Raven Press, 2015.

——. *Confederacy 101: Amazing Facts You Never Knew About America's Oldest Political Tradition.* Spring Hill, TN: Sea Raven Press,

2015.

———. *The Great Yankee Coverup: What the North Doesn't Want You to Know About Lincoln's War!* Spring Hill, TN: Sea Raven Press, 2015.

———. *Slavery 101: Amazing Facts You Never Knew About America's "Peculiar Institution."* Spring Hill, TN: Sea Raven Press, 2015.

———. *Confederate Flag Facts: What Every American Should Know About Dixie's Southern Cross.* Spring Hill, TN: Sea Raven Press, 2016.

———. *Nathan Bedford Forrest and the Ku Klux Klan: Yankee Myth, Confederate Fact.* Spring Hill, TN: Sea Raven Press, 2016.

———. *Seabrook's Bible Dictionary of Traditional and Mystical Christian Doctrines.* Spring Hill, TN: Sea Raven Press, 2016.

———. *Everything You Were Taught About African-Americans and the Civil War is Wrong, Ask a Southerner!* Spring Hill, TN: Sea Raven Press, 2016.

———. *Nathan Bedford Forrest and African-Americans: Yankee Myth, Confederate Fact.* Spring Hill, TN: Sea Raven Press, 2016.

———. *Women in Gray: A Tribute to the Ladies Who Supported the Southern Confederacy.* Spring Hill, TN: Sea Raven Press, 2016.

———. *Lincoln's War: The Real Cause, the Real Winner, the Real Loser.* Spring Hill, TN: Sea Raven Press, 2016.

———. *The Unholy Crusade: Lincoln's Legacy of Destruction in the American South.* Spring Hill, TN: Sea Raven Press, 2017.

———. *Abraham Lincoln Was a Liberal, Jefferson Davis Was a Conservative: The Missing Key to Understanding the American Civil War.* Spring Hill, TN: Sea Raven Press, 2017.

———. *All We Ask is to be Let Alone: The Southern Secession Fact Book.* Spring Hill, TN: Sea Raven Press, 2017.

———. *The Ultimate Civil War Quiz Book: How Much Do You Really Know About America's Most Misunderstood Conflict?* Spring Hill, TN: Sea Raven Press, 2017.

———. *Rise Up and Call Them Blessed: Victorian Tributes to the Confederate Soldier, 1861-1901.* Spring Hill, TN: Sea Raven Press, 2017.

———. *Victorian Confederate Poetry: The Southern Cause in Verse, 1861-1901.* Spring Hill, TN: Sea Raven Press, 2018.

———. *Confederate Monuments: Why Every American Should Honor Confederate Soldiers and Their Memorials.* Spring Hill, TN: Sea Raven Press, 2018.

———. *The God of War: Nathan Bedford Forrest as He Was Seen by His Contemporaries.* Spring Hill, TN: Sea Raven Press, 2018.

———. *The Battle of Spring Hill: Recollections of Confederate and Union Soldiers.* Spring Hill, TN: Sea Raven Press, 2018.

———. *I Rode With Forrest! Confederate Soldiers Who Served With the*

World's Greatest Cavalry Leader. Spring Hill, TN: Sea Raven Press, 2018.

———. *The Battle of Nashville: Recollections of Confederate and Union Soldiers*. Spring Hill, TN: Sea Raven Press, 2018.

———. *The Battle of Franklin: Recollections of Confederate and Union Soldiers*. Spring Hill, TN: Sea Raven Press, 2018.

———. *A Rebel Born: The Screenplay* (for the film). Written 2011. Franklin, TN: Sea Raven Press, 2020.

———. (ed.) *A Short History of the Confederate States of America* (Jefferson Davis, Belford Company, NY, 1890). A Sea Raven Press Reprint. Spring Hill, TN: Sea Raven Press, 2020.

———. (ed.) *Prison Life of Jefferson Davis: Embracing Details and Incidents in his Captivity, With Conversations on Topics of Public Interest* (John J. Craven, Sampson, Low, Son, and Marston, London, UK, 1866). A Sea Raven Press Reprint. Spring Hill, TN: Sea Raven Press, 2020.

———. *What the Confederate Flag Means to Me: Americans Speak Out in Defense of Southern Honor, Heritage, and History*. Spring Hill, TN: Sea Raven Press, 2021.

———. *Heroes of the Southern Confederacy: The Illustrated Book of Confederate Officials, Soldiers, and Civilians*. Spring Hill, TN: Sea Raven Press, 2021.

———. *Support Your Local Confederate: Wit and Humor in the Southern Confederacy*. Spring Hill, TN: Sea Raven Press, 2021.

———. *America's Three Constitutions: Complete Texts of the Articles of Confederation, Constitution of the United States of America, and Constitution of the Confederate States of America*. Spring Hill, TN: Sea Raven Press, 2021.

———. *Vintage Southern Cookbook: 2,000 Delicious Dishes From Dixie*. Spring Hill, TN: Sea Raven Press, 2021.

———. *The Bittersweet Bond: Race Relations in the Old South as Described by White and Black Southerners*. Spring Hill, TN: Sea Raven Press, 2022.

———. (ed.) *The Rise and Fall of the Confederate Government* (Jefferson Davis, D. Appleton, New York, 1881). 2 vols. A Sea Raven Press Facsimile Reprint. Spring Hill, TN: Sea Raven Press, 2022.

———. *Secrets of Celebrity Surnames: An Onomastic Dictionary of Famous People*. Cody, WY: Sea Raven Press, 2023.

———. *I, Confederate: Why Dixie Seceded and Fought in the Words of Southern Soldiers*. Spring Hill, TN: Sea Raven Press, 2023.

———. *Twelve Years in Hell: Victorian Southerners Expose the Myth of Reconstruction, 1865-1877*. Cody, WY: Sea Raven Press, 2023.

———. *Seabrook's Complete Battle Book: The War Between the States, 1861-1865.* Cody, WY: Sea Raven Press, 2023.

———. *The Hampton Roads Conference: The Southern View.* Cody, WY: Sea Raven Press, 2024.

———. *Rocky Mountain Equines: A Photographic Collection of Horses, Donkeys, and Mules of the American West.* Cody, WY: Sea Raven Press, 2024.

———. *Rocky Mountain Bison: A Photographic Collection of Bison of the American West.* Cody, WY: Sea Raven Press, 2024.

———. *Mysterious Invaders: Twelve Famous 20th-Century Scientists Confront the UFO Phenomenon.* Cody, WY: Sea Raven Press, 2024.

———. *We Called Him Jeb: James Ewell Brown Stuart as He Was Seen by His Contemporaries.* Cody, WY: Sea Raven Press, 2024.

———. *Your Soul Lives Forever: Documented Victorian Case Studies Proving Consciousness Survives Death.* Cody, WY: Sea Raven Press, 2024.

———. *Authentic Victorian Ghost Stories: Genuine Early Reports of Apparitions, Wraiths, Poltergeists, and Haunted Houses.* Cody, WY: Sea Raven Press, 2024.

———. *The Greatest Jesus Mystery of All Time: Where Was Christ Between the Ages of 12 and 30?* Cody, WY: Sea Raven Press, 2024.

———. *Vitamin D: The Miracle Treatment for Nearly Every Disease and Health Issue.* Cody, WY: Sea Raven Press, 2024.

———. *Manmade: Male Inventors Who Created the Modern* World. Cody, WY: Sea Raven Press, 2025.

———. *Jesus and the Gospel of Thomas: A Christian Mystic's View of Christianity's Most Important Ancient Text.* Cody, WY: Sea Raven Press, 2025.

———. *The Hunter-Gatherer Principle: Evolutionary Biology and the Case for Sex-Based Female Sports.* Cody, WY: Sea Raven Press, 2025.

———. *The Way of Holiness: The Story of Religion and Mythology, from the Cave Bear Cult to Christianity: A Study of the Origins, Development, Functions, Symbols, and Themes of Spiritual Thought.* Unpublished manuscript.

———. *Mothers and Bachelors: Ending the Battle of the Sexes—A New Approach to Marriage and the Family Based on the Sciences of Anthropology, Primatology, and Sociobiology.* Unpublished manuscript.

———. *Seabrook's Encyclopedia of Religion and Myth.* Unpublished manuscript.

———. *Families Around the World: A Children's Guidebook to the Marriages and Families of Different Cultures.* Unpublished manuscript.

———. *The True Legend of King Arthur: The Magical Story of Britain's Most Famous Ruler.* Unpublished manuscript.

———. *Glimpses of Heaven: A Guidebook to the Near-Death Experience.* Unpublished manuscript.

Sereno, Paul C. *Dragons of the Dust: The Paleobiology of the Cretaceous Desert.* New Brunswick, NJ: Rutgers University Press, 1994.

Sinibaldi, Robert W. *Ice Age Florida: In Story and Art.* Pittsburgh, PA: Dorrance Publishing Co., 2021.

Stark, Mike. *Chasing the Ghost Bear: On the Trail of America's Lost Super Beast.* Lincoln, NE: University of Nebraska Press, 2022.

Sternberg, Charles H. *The Life of a Fossil Hunter.* New York: Henry Holt and Company, 1909.

Stewart, Anthony J. *Vanished Giants: The Lost World of the Ice Age.* Chicago, IL: University of Illinois Press, 2021.

Sues, Hans-Dieter. Sues, Hans-Dieter. *The Carnivorous Dinosaurs.* Bloomington, IN: Indiana University Press, 2005.

———. *Triassic Life on Land: The Great Transition.* New York: Columbia University Press, 2010.

Switek, Brian. *Written in Stone: Evolution, the Fossil Record, and Our Place in Nature.* New York: Bellevue Literary Press, 2010.

Teichert, Curt. *A Paleontological Life: The Personal Memoirs of Curt Teichert.* Ithaca, NY: Cornell University Press, 2012.

The American Journal of Science and Arts. Various issues. New Haven, CT: Silliman and Dana, 1840s.

Thompson, David A. W. *A Glimpse of Eden: The Teaching of Paleontology.* London: Cambridge University Press, 1965.

Ulrich, E. O., August F. Foerste, A. K. Miller, and A. G. Unklesbay. *Ozarkian and Canadian Cephalopods: Part III: Longicones and Summary.* Boulder, CO: Geological Society of America, 1944.

Weishampel, David B., Peter Dodson, and Halszka Osmólska. *The Dinosauria* (second edition). Berkeley: University of California Press, 1990.

Witton, Mark P. *Pterosaurs: Natural History, Evolution, Anatomy.* Princeton, NJ: Princeton University Press, 2013.

———. *King Tyrant: A Natural History of Tyrannosaurus rex.* Princeton, NJ: Princeton University Press, 2025.

Zanno, Lindsay E. *Dinosaurs: A Concise Natural History.* Baltimore: Johns Hopkins University Press, 2015.

———. *The Theropods: Dinosaur Facts and Figures.* Princeton, NJ: Princeton University Press, 2024.

INDEX

MEET THE AUTHOR

LOCHLAINN SEABROOK is a prolific lifelong researcher, historian, author, naturalist, artist, and composer whose knowledge and experience span numerous fields. His remarkable productivity stems from his broad interests, decades of meticulous research, and an unwavering daily devotion to writing and creative exploration.

The idea of specializing in a single subject is a modern invention. In the spirit of the great polymaths—Aristotle, Isaac Newton, Benjamin Franklin, and Thomas Jefferson—Seabrook works across dozens of disciplines, with intellectual pursuits encompassing history, science, philosophy, religion, and the arts. The result is an expansive body of original writings that distill years of careful analysis into clear, accessible language for the general reader.

Rejecting the narrow confines of modern specialization, Seabrook views all knowledge as intrinsically interconnected. This integrative vision, combined with long hours of focused, solitary study and a rigorous work ethic, has enabled him to produce an extraordinary corpus of literature uniting the sciences and the humanities—a natural outgrowth of a lifetime devoted to inquiry, creativity, and the preservation of evidence-based history.

AMERICAN POLYMATH LOCHLAINN SEABROOK is a bestselling author, award-winning historian, and acclaimed multidisciplinary artist. A descendant of the families of Alexander Hamilton Stephens, John Singleton Mosby, Edmund Winchester Rucker, and William Giles Harding, the neo-Victorian scholar is a 7th generation Kentuckian, and one of the most prolific and widely read traditional writers in the world today. Known by literary critics as the "new Shelby Foote," the "American Robert Graves," the "Southern Joseph Campbell," and the "Rocky Mountain Richard Jefferies," and by his fans as the "the best author ever," he is a recipient of the United Daughters of the Confederacy's prestigious Jefferson Davis Historical Gold Medal, and is considered the foremost Southern interpreter of American Civil War history—or what he refers to as the War for the Constitution (1861-1865).

A lifelong litterateur, the Sons of Confederate Veterans member has authored and edited books ranging in topics from ancient and modern history, politics, science, comparative religion, diet and nutrition, spirituality, astronomy, entertainment, military, biography, mysticism, anthropology, cryptozoology, photography, and Bible studies, to natural history, technology, paleography, music, humor, gastronomy, etymology, paleontology, onomastics, mysteries, alternative health and fitness, wildlife, alternate history, comparative mythology, genealogy, Christian history, and the paranormal; books that his readers describe as "game changers," "transformative," and "life altering."

One of America's most popular living historians, nature writers, autodidacts, and Transcendentalists, he is a 17th generation Southerner of Appalachian heritage who descends from dozens of patriotic Revolutionary War soldiers and Confederate soldiers from Kentucky, Tennessee, North Carolina, and Virginia. Also a history, wildlife, and nature preservationist, the well-respected scrivener began life as a child prodigy, later maturing into an archetypal Renaissance Man and classical polymath.

Besides being cofounder and co-CEO of Sea Raven Press, an accomplished writer, author, historian, biographer, lexicographer, encyclopedist, neologist, publisher, editor, poet, polymathic creative, onomastician, etymologist, and Bible authority, the influential prosateur is also a Kentucky Colonel, eagle scout, entrepreneur, businessman, composer, screenwriter, nature, wildlife, and landscape photographer, videographer, and filmmaker, artist, artisan, painter, watercolorist, sculptor, ceramic artist, visual artist, sketch artist, pen and ink artist, graphic artist, graphic designer, book designer, book formatter, editorial designer, book cover designer, publishing designer, Web designer, poster artist, digital artist, cartoonist, content creator, inventor, aquarist, genealogist, ufologist, jewelry designer, jewelry maker, former history museum docent, teacher's assistant, and a former Red Cross certified lifeguard, ranch hand, zookeeper, and wrangler. A contemporary songwriter (of some 3,000 songs in a dozen genres), he is also a pianist, organist, drummer, bass player, rhythm guitarist, rhythm mandolinist, percussionist, electronic musician, synthesist, clavichordist, harpsichordist, classical composer, jingle composer, film composer (currently his musical work has been featured in 11 movies), lyricist, band leader, multi-instrument musician, lead vocalist, backup vocalist, session player, music producer, and recording studio mixing engineer, who has worked and performed with some of Nashville's top musicians and singers.

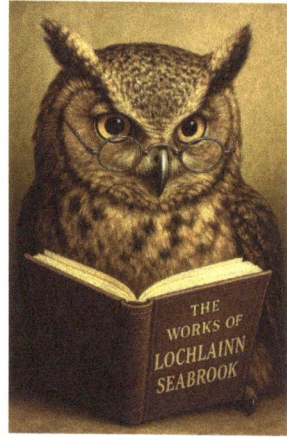

Currently Seabrook is the multi-genre author and editor of over 100 adult and children's books (totaling some 30,000 pages and 15,000,000 words) that have earned him accolades from around the globe. His works, which have sold on every continent except Antarctica, have introduced hundreds of thousands to vital facts that have been left out of our mainstream books. He has been endorsed internationally by leading experts, museum curators, award-winning historians, chart-topping authors, celebrities, filmmakers, noted scientists, well regarded educators, TV show hosts and producers, renowned military artists, venerable heritage organizations, and distinguished academicians of all races, creeds, and colors.

He currently holds two interesting world records: He is the author of the most books on American military officer Nathan Bedford Forrest, and he was the first to publicize and describe the 19th-Century platform reversal of America's two main political parties, namely that Civil War era Democrats (primarily in the South—the Confederacy) were Conservatives, while Civil War era Republicans (primarily in the North—the Union) were Liberals.

Of northern, western, and central European ancestry, he is the 6th great-grandson of the Earl of Oxford and a descendant of European royalty through his Kentucky father and West Virginia mother. A proud descendant of Appalachian coal miners, trainmen, mountain folk, and wilderness pioneers, his modern day cousins include: Johnny Cash, Elvis Presley, Lisa Marie Presley, Billy Ray and Miley Cyrus, Patty Loveless, Tim McGraw, Lee Ann Womack, Dolly Parton, Pat Boone, Naomi, Wynonna, and Ashley Judd, Ricky Skaggs, the Sunshine Sisters, Martha Carson, Chet Atkins, Patrick J. Buchanan, Cindy Crawford, Bertram Thomas Combs (Kentucky's 50th governor), Edith Bolling (second wife of President Woodrow Wilson), Andy Griffith, Riley Keough, George C. Scott, Robert Duvall, Reese Witherspoon, Lee Marvin, Rebecca Gayheart, and Tom Cruise.

A constitutionalist, avid outdoorsman, wilderness conservationist, and gun rights advocate, Seabrook is the author of the international blockbuster, *Everything You Were Taught About the Civil War is Wrong, Ask a Southerner!* He lives with his wife and family in the magnificent Rocky Mountains, heart of the American West, where you will find him writing, hiking, and filming.

For more information on Mr. Seabrook visit

LochlainnSeabrook.com

Praise for Author-Historian-Artist
Lochlainn Seabrook

Comments from our readers around the world

★ "Lochlainn Seabrook is a genius writer!" — STEVEN WARD

★ "Best author ever." — EMILY (last name withheld)

★ "We get asked a lot what books we use and read. We don't do many modern historians, but we make an exception for some, and Lochlainn Seabrook is one of them. His works are completely well researched from original documents, and heavily footnoted and documented." — SOUTHERN HISTORICAL SOCIETY

★ "Looking forward to more Lochlainn Seabrook books, my favourite historian!" — ALBERTO IGLESIAS

★ "Lochlainn Seabrook is one of the finest authors on true history in this century. His books should be on every student's desk." — RONDA SAMMONS RÉNO

★ "All of Col. Seabrook's books are great. I have bought most of them and want to end up buying them all." — DAVID VAUGHN

★ "Lochlainn pulls together such arcane facts with relative ease, compiling these into ordinary prose that strike to the heart with substance, no fluff-speak. I am awestruck! Really. He is an inspiration to me. . . . He is truly a revolutionist. He dares to speak what others whisper; he writes with a boldness and an authoritative knowledge that is second to none." — JAY KRUIZENGA

★ "Mr. Lochlainn Seabrook is . . . the most well researched and heavily documented author I've ever read. His books are must haves. Everything he writes should be required reading! I assure you, you won't be disappointed. One simply cannot go wrong with his books. Mr. Seabrook is awesome! . . . I have never read any other author as well researched and footnoted as him. I've been in love with Mr. Seabrook for almost 5 years now. His quick wit and logic is enough reason to purchase his books. But the mere fact that he's so extensively researched is icing on the cake. Mr. Seabrook is my favorite, hands down." — LANI BURNETTE RINKEL

★ "My favorite book is the Bible. Lochlainn Seabrook wrote my second favorite book." — RICHARD FINGER

★ "I have a new favorite author and his name is Lochlainn Seabrook." — J. EWING

★ "Lochlainn Seabrook is an incredible writer and I love all of his books on the South. . . . His writing is brilliant. . . . I look forward to reading more of his masterpieces. Thank you." — JOEY (last name withheld)

★ "It's hard to choose just one of Lochlainn's books!" — ROSANNE STEELE

★ "Mr. Seabrook, thank you ever so much for blessing us with your most enlightening works." — LAURENCE DRURY

★ "I recommend anything written by Lochlainn Seabrook." — HOTRODMOB

★ "Awesome books . . . by a great writer of truth, Lochlainn. Thank you so much. Keep up the great work you do." — WILDBUNCH19INF

★ "I love Lochlainn Seabrook's style and approach. It's not the 'norm.' What a miracle his books are. . . . He is a literal life changing author! Amazing books!" — KEITH PARISH

★ "I adore Mr. Seabrook's style and I love his books. I love an author that does proper research, and still finds a way to engage the reader. Mr. Seabrook does an admirable job of both." — DONALD CAUL

★ "Lochlainn Seabrook's books are much more well researched and authoritative than those eminently celebrated as being the authorities on the subjects he writes on. You can always trust to find the truth in his writings. . . . He does not rewrite history, but instead shows it as it is." — GARY STIER

★ "I love all of Colonel Seabrook's books. They are informative and enlightening, and his warm Southern hospitality writing style makes you feel right at home." — KEITH CRAVEN

★ "Lochlainn Seabrook's work is an absolute treasure of scholarship and historic scope." — MARK WAYNE CUNNINGHAM

★ "Mr. Seabrook's command of . . . history is breathtaking. . . . He deserves great renown—check out his books!" — MARGARET SIMMONS

★ "I love Seabrook's writings. LOVE!!! . . . So grateful to know the truth! Keep writing Lochlainn!!!" — REBECCA DALRYMPLE

★ "Lochlainn Seabrook . . . [has] probably [written] the best book on mental science in existence by a living author. Along with Thomas Troward, Emmet Fox, and Jack Addington, Mr. Seabrook is one of the top four mental science authors of all time, since biblical times." - IAN BARTON STEWART

★ "Glad I discovered Mr. Seabrook! . . . He writes eye opening books! Unbelievable the facts he unearths - and he backs it all up with truth, notes, footnotes, and bibliography! . . . He always amazes me! His books always see the whole picture. His timelines and bibliographies are incredible. He always provides carefully reasoned arguments! He's the best. To me I think he's better than the late great Shelby Foote! America needs more like Lochlainn Seabrook. I can't wait to own all of his books on the war someday. Everyone who wants the Truth, who seeks the Truth and wants the full story, should read his books." — JOHN BULL BADER

★ "I love all of Colonel Seabrook's books!" — DEBBIE SIDLE

★ "Lochlainn Seabrook is well educated and versed in what he writes and I'm impressed with the delivery." — THOMAS L. WHITE

★ "Lochlainn Seabrook is the author of great works of scholarship." — JOHN B. (last name withheld)

★ "Thank you Lochlainn Seabrook for your wonderful books! You are the real deal! You are an amazing author and I love your books!!" — SOPHIA MEOW CELLIST

★ "I really enjoy Mr. Seabrook's books! His knowledge is beyond belief!" — SANDRA FISH

★ "Love Lochlainn Seabrook. Awesome!!" — ROBIN HENDERSON ARISTIDES

★ "Kudos to Lochlainn Seabrook who is a very good and informative professional truthful historian. We need more like him!" — AMY VACHON

Nurture Your Mind, Body, and Spirit!

READ THE BOOKS OF

SEA RAVEN PRESS

Visit our Webstore for a wide selection of wholesome, family-friendly, evidence-based, educational books for all ages. You'll be glad you did!

Artisan-Crafted Books & Merch From the Rocky Mountains

THANK YOU FOR SUPPORTING OUR SMALL AMERICAN FAMILY BUSINESS!

SeaRavenPress.com

Visit our sister sites:
LochlainnSeabrook.com
YouTube.com/user/SeaRavenPress
YouTube.com/@SeabrookFilms
Rumble.com/user/SeaRavenPress
Pond5.com/artist/LochlainnSeabrook

If you enjoyed this book you may be interested in some of Colonel Seabrook's popular related titles:

☛ NORTH AMERICA'S AMAZING MAMMALS: AN ENCYCLOPEDIA FOR THE WHOLE FAMILY
☛ THE CONCISE BOOK OF OWLS: A GUIDE TO NATURE'S MOST MYSTERIOUS BIRDS
☛ THE CONCISE BOOK OF TIGERS: A GUIDE TO NATURE'S MOST REMARKABLE CATS
☛ ROCKY MOUNTAIN EQUINES: A PHOTOGRAPHIC COLLECTION OF HORSES, DONKEYS, AND MULES OF THE AMERICAN WEST
☛ ROCKY MOUNTAIN BISON: A PHOTOGRAPHIC COLLECTION OF BISON OF THE AMERICAN WEST

Available from Sea Raven Press and wherever fine books are sold

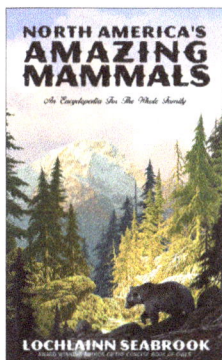

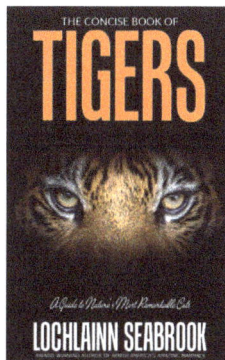

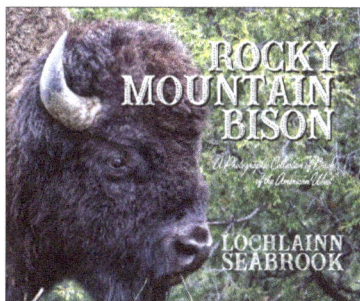

SeaRavenPress.com

www.ingramcontent.com/pod-product-compliance
Lightning Source LLC
Chambersburg PA
CBHW041732200326

41518CB00019B/2578